MONETIZING 21ST CENTURY TELEVISION

How to Profit in the Coming TV Revolution

Frank A. Aycock, Ph.D.

ISBN-10: 1540760340
ISBN-13: 9781540760340

To my beautiful wife, Gail,
My inspiration for all things,
My cheerleader, and the love of my life;
To my two sons and their wives,
Who will enjoy all that 21st Century Television
Has to offer; and to my wonderful grandson, Elias,
For whom 21st Century Television will be old hat,
And who will enjoy the dawning of
22nd Century Television

TABLE OF CONTENTS

ACKNOWLEDGEMENTS

Writing a book is a solitary endeavor and yet it isn't. While the author writes the words, there are numerous others without whom the author might never finish the book. This book, especially, was a work a long time in the making. After producing three books in four years, this one has taken more than three years to finish. Trying to properly and correctly write about monetizing the coming television universe is not something I took lightly – it's too important to all those who are part of the 21st Century Television universe. Individual careers and whole segments of the 21st Century Television industries will live or die on whether or not they can transition from 20th Century Television monetization to 21st Century Television monetization. This book is designed to do just that – create a win-win-win scenario for everyone.

I would like to thank my good friend and colleague, Ms. Laura Brittain for her assistance. Laura has a creative genius about her that produced the amazing cover for this book. She is a wonderful creative professional and an amazing person, and I truly appreciate all she has done to make this book possible.

I cannot thank enough Jeanette Watson and the team at the Deloitte Corporation who once again so graciously provided me with the many tables and data that I used in the chapter on the viewers. Without their diligent efforts and willingness to share their work, the story of the changing television viewer could not have been told in a timely manner. They have been wonderful to work with.

I want to extend a special note of thanks to Joe Mandese, the editor in chief of *MediaPost,* for his permission to use an extensive portion of the article "How Artificial Intelligence Ties Into Programmatic Media." The article provided a real insight for the chapter on Programmatic Advertising, and I appreciate his willingness to permit me to use the article.

Likewise, I want to thank Patrick McKenna, CEO of Strikesocial. com, for his permission to use extensive portions of the article, "The differences between AI, machine learning, programmatic buying and deep learning." It, too, was a crucial part of the chapter on Programmatic Advertising, and I do appreciate Patrick's agreement to allow me to use the article in depth.

I want to thank the KDP publishing team, as well as Amazon and its founder and CEO, Jeff Bezos, for making it possible for an author writing on the future of television to publish with the future of book publishing. Most of the members of both organizations will never know me, more than likely, and I doubt I will ever meet Jeff Bezos, but I can say that it is a pleasure to publish in this manner.

As every author knows, you just don't sit down one day and bang out 60,000 or so words of a manuscript. Months and months of research are necessary. Fortunately, I have a number of great researchers as part of my team – yet they don't even know that they are. So I want to say thank you to all the folks who publish *Digital Media Wire; VideoNuze;* Shelly Palmer of *Shelly Palmer Digital Living;* and the people whose work at the various Smartbriefs, *TVNewsCheck,* and other such aggregators for providing the background research that made it possible for me to have the latest information at my fingertips every day. They are all very important, and without them I don't know how I would be able to write this book, as well as my previous books, as their constant searching for and delivering the latest information in this amazing time we live in keeps me energized and excited about pursuing all the myriad aspects of this television revolution that I call 21st Century Television. I cannot thank them enough.

Finally, and most importantly, I want to thank my two sons and my daughter-in-laws for their encouragement and support; my amazing grandson who inspires me to continue to look ahead and try to understand the television world he will enjoy; and Gail, my beautiful wife and the love of my life for 38 years, without whom I know this book, as well as all my others before and the ones to come, would be languishing only in my mind. She has been my support, my cheerleader, my copyeditor, and the one who convinced me that a book about the future of television should be published in the future style of publishing. She has been right beside me every step of the way, and she believes in all my books and in me. I am truly a blessed man.

INTRODUCTION

In his opening keynote address to the 2008 National Association of Television Program Executives convention, NBC Universal president and CEO Jeff Zucker uttered that now-legendary statement "Our challenge with all these ventures is to effectively monetize them so that we do not end up trading analog dollars for digital pennies."[1] He added, "This is the No. 1 challenge for everyone in this industry today."[2] In this famous statement, Zucker lifted the hearts of every broadcaster in the room, and set back the movement toward 21st Century Television by who knows how long. To be fair, Zucker has changed his view – slightly. He now says the statement as "analog dollars to digital quarters," recognizing that the online world of video is beginning to make more revenue, but evidently, Zucker still does not like the idea of change.[3]

Zucker was making the point that, in his view, moving from existing television to 21st Century Television (not just simply moving from analog to digital television) was a recipe for disaster – that television could only stay profitable if it dug its heels in and continued to insist on doing things the "same old way." Putting television on the Internet over broadband, Wi-Fi, or cellular simply was not economically feasible – that television would be trading analog advertising dollars for digital advertising pennies because viewers would not watch advertising over the Internet. Television had to remain as it was and had been. To do otherwise would be the death of television.

It's now been more than a decade since Zucker first uttered his famous remark, and, for virtually all that time, the notion has been fondly held by broadcasters that television cannot be delivered through the Internet because, as everyone knows, "people watch television over the Internet because they want to avoid all the commercials that are aired on over-the-air and cable television." It is truly amazing how publicly devoted to the erroneous mantra first uttered by Zucker the broadcasters have been since that time. The key word in the last sentence is "publicly" because that notion is slowly (and often quietly) changing as new companies develop and make available new tools for monetizing 21st Century Television. While the local broadcast media have been the most likely to publicly cling to this archaic notion, the cable, satellite, and even the broadcast networks are looking for answers to monetizing their products. Whether it be ATSC 3.0, expansive retransmission consent agreements, or 21st Century Television advertising, much of the legacy media continue to look for ways to improve their bottom lines.

Since the late 1940s, American viewers have been continually conditioned to the idea that television will have commercial breaks in the programming. Initially, commercials generally ran for one minute at a time. As television advertising progressed and viewer sophistication with television advertising grew, the one minute spot gave way to the classic 30-second spot, which dominated the television advertising landscape for most of the history of the industry. Even at this time, the 30-second spot lives on, although – once again because of increasing viewer sophistication with television advertising – more 20-second and even 15-second spots are airing on programs. It is now possible to tell the advertising story in as little as one-fourth the time as it took to tell the story when television was young, thanks to the ever-increasing sophistication of both the advertising creative professionals and the viewing public.

Why, then, have broadcasters been so certain that the public will not watch commercials on television programs viewed over the Internet? After all, the broadcasters and advertising professionals have worked so hard and so diligently for more than half a century to make sure the viewing public is conditioned to watching as many as 64 commercials an hour. (Assuming all commercials during the hour are the 15-second variety – 64 commercials would be the maximum number that could run.) Why buy into a notion that is completely illogical and counter to everything that the broadcasters and advertising professionals have worked so hard to perfect since the late 1940s? It makes no sense, and the longer broadcasters hold to the notion that viewers will not watch commercials (or minimal numbers of commercials at the most) in programming delivered over the Internet, the more they reinforce that notion until it becomes a self-fulfilling prophecy.

Instead, broadcasters and all legacy media content providers should embrace the Internet for the delivery system it is, and make good use of the unique capabilities for advertising that the Internet provides. Viewers conditioned to watching television in the traditional ways will watch television the same way over the 21st Century Television delivery system of the Internet. Younger viewers must also be conditioned to watch commercialized television over the Internet. Otherwise, they will be the generation to begin the self-fulfilling prophecy of advertising as unacceptable on television. Whether the content providers like it or not, television will move to the new delivery system of IPTV and commercials will continue to air in programs as they have. If not, Jeff Zucker will be right. If he ultimately is, it will be the content providers' fault, not the viewers', and it will be the content providers who will be the ones that pay the ultimate price.

So how will advertising work on 21st Century Television? In reality, it will be more efficient and more effective than on television today. Traditional television advertising has been a shotgun approach – you advertised to the largest number of people and hoped that a reasonable number of viewers responded to the advertisement. If they did, the ad was effective. If not, well, you tried something else. While there is always talk about the target audience, the target market, and advertising to specific demographic and psychographic groups, in reality, advertising on television still remains a shotgun approach because it is simply not possible to target more narrowly on traditional over-the-air, broadcast television.

For cable and direct-to-home satellite services, it is easier to target specific groups because so many of the channels are niche-audience focused. However, even on those most niche-oriented channels, targeted advertising is not nearly as specific as it could be. Most cable- and satellite-delivered channels see their viewers as aggregate audiences even if they are more tightly defined, at least on their traditional cable delivery systems. While, say, the History Channel may target history buffs, or ESPN might target football fans in the fall, basketball fans in the winter and early spring and baseball fans in the summer, those viewers are still lumped into single aggregate groups and only rudimentarily broken out any further. Nevertheless, the examples of the niche audiences on cable and satellite TV today begin to show the possibilities for 21st Century Television. Because the programming is already niche, the advertising can also be somewhat niche – but not to the degree 21st Century Television will offer.

Then there is the question of interactivity. Traditional television is simply one-way television. Television advertisements are delivered to the audience with almost no way for the audience to respond directly to the message. The communication is one way, not two way. Cable has the ability to deliver interactive, two-way advertising, but, for the most part, it has not done so. Traditionally, cable television channels have taken the same approach that the broadcast television channels have taken – delivering their advertisements in a one-way manner.

Whether it is the broadcast channels or the cable specific channels, one way advertising – as well as one-way programming to their audiences – is the time-honored tradition. The two-way programming that has occurred on television and more often on cable is in the form of home shopping networks, infomercials, and hard-sell advertisements that request immediate 1-800 telephone call orders, or in the form of certain reality programs where the audience is urged to call, text, or log on to the channel's website to vote for a favorite participant. While these are all – in a sense – interactive, in reality, they are a poor beginning for what is possible with true, interactive television advertising.

At the time of the first edition of the first book in this series – *21st Century Television: The Players, The Viewers, The Money* – some of the more forward-thinking cable system owners, especially those such as Comcast and Time Warner, were beginning experiments with interactive advertising.[4] Cablevision added an opt-in interactive option to its cable services called Optimum Select, where viewers could request coupons, more information about a product or service, or other longer-form content.[5]

Twenty-first Century Television, delivered as IPTV, provides a whole new level of advertising possibilities that are either not available, or that are more available, than what advertising is currently on legacy television. In fact, some have suggested that we could see not only completely interactive television commercials, but also true microadvertisements, or advertisements literally targeted to an individual. The first is certainly possible; the second is not economically feasible for the reason that it is not possible to make millions of different commercials to deliver to millions of different viewers on a true, one-to-one basis. However, an alternate form of advertising, using certain aspects of microadvertising, will be economically feasible as well as extremely desirable, and will be discussed later in this book.

Total interactivity will be a mainstay of IPTV. With a click of a remote control or a mouse – or even a voice command or a wave of the hand – viewers will be able to change views of the product, change colors, open windows to provide additional product information, find locations to purchase the product, etc. Additionally, depending on the specific product, interactivity will include the ability to make an immediate purchase, upgrade an existing product, extend warranties, etc. Twenty-first Century Television, because it is delivered over IPTV (whether through wired connection, Wi-Fi, mobile, or other delivery avenues), will operate much like what is already being delivered on a trial basis over computer websites today. With the connected televisions that set manufacturers provide to users, with separate windows on the screen for widgets, chat rooms, websites, etc., the ability to have total interactivity is available to advertisers. They just have to deliver the commercials that have interactive elements in them, so that the viewer can make decisions immediately. Connected television sets are currently available from virtually all of the television set manufacturers, complete with software from Rovi, Google, Roku, and other software makers that make the multi-screen, interactive television sets possible.[6] Compared to the sets earlier this decade, the latest connected television sets are even more impressive than their predecessors. Even so, they are still extremely basic to the television sets of the next few years and into the future. Nevertheless, the hardware and software are moving closer to what will be necessary to produce the interactive television set of the future. Such interactivity will provide not only a much richer purchasing environment over the television, resulting in more impulse sales of products and services, but will also provide the advertising agencies and their clients with continuous feedback on the success or failure of their advertising campaign creatives.

What is most important to advertising agencies and their clients, however, will be the ability to target much more narrowly their preferred audiences and markets. The ability to narrowly target these specific audiences will allow advertising professionals and their clients to have maximum impact with their commercials and maximum success with their campaigns. Targeting the audiences will be much more specific than even what occurs on cable and satellite niche programming today. A good term for this first step in a three step process of advertising is Aggregated Targeted Microadvertising, or ATMA. The author chose the phrase aggregated targeted micro-advertising because at the time there was no one term generally accepted by the television or advertising industries.[7]

Since at least 2015, there has been such a term – dynamic ad insertion, or DAI. However, to use the term DAI to refer to the entire three step process is incorrect and short-sighted. According to the Interactive Advertising Bureau, dynamic ad insertion is "The process by which an ad is inserted into a page in response to a user's request. . . .At its simplest, dynamic ad placement allows for multiple ads to be rotated through one or more spaces. In more sophisticated examples, the ad placement could be affected by demographic data or usage history for the current user."[8] The Interactive Television Institute's *itv dictionary,* says about dynamic ad insertion, "With dynamic ad insertion, network operators can provide targeted ads that can be swapped in and out of that tele-vision program as it's delivered to the enduser."[9] Other terms for DAI include *addressable advertising* and *targeted advertising,* although neither term is correct as a synonym for DAI as both terms describe the full three step process of 21st Century Television advertising, while DAI only describes the third step in the process.

While DAI describes the end result of the action of inserting a variety of ads into a program's particular availability or designated spot ad time, based upon the viewer's preferences, the three-step process (discussed later in this book) of ATMA, Programmatic Media Buying, and DAI provides a better description of the process by which the end result is accomplished. However, the author does acknowledge that dynamic ad insertion is the current "buzz" term used to describe the three-step process of addressable advertising, even though DAI is, in reality, only the end result of the process.

ATMA is not specifically microadvertising. Microadvertising is advertising that directs a different ad to each person, so for one million people an advertiser would need one million different advertisements. In ATMA, the key word is "aggregated." While microadvertising has been thought to be neither feasible nor economically viable, aggregated targeted microadvertising will, most likely, be the way of the future. Early generations of ATMA were being developed and introduced to television as of mid-2012, and today, there is a strong, but still fledgling DAI industry for 21st Century Television. The major impediment to full adoption is the concern by the legacy media to make the changes in delivery systems necessary for DAI to be successful on a macro scale.

ATMA makes possible the delivery of advertising to specific individuals much like microadvertising, but, unlike microadvertising, it uses only a set number of advertisements and directs one of those several advertisements to each specific individual in the targeted subset of the audience. There continue to be small-scale (comparative to traditional advertising methods) usage of DAI on websites like Hulu and the legacy network websites such as NBC, where they provide the viewer with two advertisements and let them choose which of the two ads they wish to watch. As far back as February 4, 2010, an article in the *Wall Street Journal* reported that research demonstrated that when viewers were allowed to choose from among three advertisements which one they wanted to watch, the consumers were more likely to watch the advertisement, recall the advertisement they watched, and were more engaged with the advertisement. The research study, conducted by a group including the advertising agency Publicis Groupe, Microsoft, Yahoo, CBS, and Hulu, also showed that choice was preferred by consumers over approximately 30 other forms of advertising formats currently in use.[10] While the results showed choice to be preferred even over interactive online videos and clickable videos, it must be remembered that only the choice format allowed the consumer to have his/her preferred advertisement.

This book is divided into three parts. Part One first looks at a short history of television advertising in an effort to set the stage for successful monetization of 21ˢᵗ Century Television. It's important to see where the industry has come from to understand where the industry must go to continue being successful. The book then turns to the different industries that are most likely to need and/or make use of 21ˢᵗ Century Television monetization to successfully compete in the new universe of television. Finally, this section focuses on the changing television viewer with the major focus on those generations that will be at the forefront of 21ˢᵗ Century Television – Generation X, Millennials, and Generation Z. It is important to understand these generations to understand how to reach them with new types of traditional monetization methods that today are anathemas to those young, Netflix-watching viewers.

The first section of Part Two focuses on a three-step process for delivering advertising to 21ˢᵗ Century Television viewers. Aggregated Targeted Microadvertising (ATMA) is the first step; Programmatic Advertising (shortened form: Programmatic) – or, more correctly, Programmatic Media Buying – is the second; and Dynamic Ad Insertion (DAI) is the third and final step in the process. While both Programmatic and DAI have been well known and well discussed in the last few years, the antecedent, ATMA, in any discussion is simply nonexistent, or at best, the process is subsumed under one or the other two terms. However, to do so minimizes the importance of accomplishing ATMA first, before moving to Programmatic and DAI.

So, because advertising on 21ˢᵗ Century Television is/will be a three step-process, the book will discuss first, ATMA; second, Programmatic Media Buying; and, third, Dynamic Ad Insertion. In doing so, the author hopes to make clear that all three are needed to fully accomplish the monetization of 21ˢᵗ Century Television – at least from the advertising part of that monetization.

The second section of Part Two looks at other aspects of monetizing 21ˢᵗ Century Television that can and will be used along with advertising. This section first looks at a different type of traditional monetization that must undergo a transformation - product placement. The chapter on product placement is centered around the introduction of a new concept – "Ubiquitous Product Placement," or UPP – an expansion of the long-used and well-known product placement that has been a part of television practically since its beginning, and a part of film before that. The advanced technology available to today's 21st Century Television professionals makes it possible to expand the capabilities of UPP to reach audiences in ways not possible before, and makes it possible to combine UPP with 21st Century Television advertising to form a powerful driving force for monetizing 21st Century Television, making possible levels of revenue only dreamed about until now.

In addition to 21st Century Television advertising and ubiquitous product placement, the second section of Part Two also looks at how promotion plays a critical part in this new television universe. While promotion is the one area where today's television executives are hard at work, the chapter on promotion also offers new aspects and ways of reaching the audience for consideration.

Finally, the second section of Part Two takes a look at what global expansion of 21st Century Television throughout the world will mean for advertising, product placement, and promotion to the different television industries. Television in the 21st Century will become a truly global commodity, with audiences around the world enjoying whatever television they want to watch, when they want to watch, where they want to watch, and on whatever platform they choose to watch, without having to consider how national borders impact their viewing.

Part Three will bring together all the various aspects of 21st Century Television monetization and demonstrate how the different aspects will work together. As a conclusion, then, Part Three will finish with a scenario of the future of 21st Century Television monetization, and how it will lead to a new "Diamond Age" of television, far surpassing any "Golden Age" one might consider.

PART I:
TRADITIONS

CHAPTER 1
A QUICK HISTORY OF TV ADVERTISING

Since the beginning of television in the 1940s and 1950s, advertising has been the driving force behind the revenues that private TV stations have depended on to fund their costs and provide profits. Today, there are very few countries where TV stations, including the state-run-or-supported TV stations/networks, don't have advertising as a main or major source of revenue. Also today, many, if not most, countries of the world have privately-run, advertiser-driven television stations that compete directly with state run-or-supported television.

This competition between the private and state run/supported television stations is often not a happy one because of the ability of the state television to undercut the pricing of the private stations. State run/supported television stations can undercut the advertising prices of the private stations because they also receive funding from the government, usually in the form of a line-item in the federal budget[1] or as a tax[2] that citizens are required to pay. Depending on the country, the competition can be fierce between the private and state run-or-supported television stations for programming and advertising. For private stations around the world, and those public/state television stations that do advertise, the television advertising vehicle is the commercial.

Traditionally, the television commercial around the world has followed a relatively constant pattern, depending on the length of time a country's television industry has been producing and airing commercials. The pattern will often also be impacted by whether or not, and for how long, the country's radio industry has aired commercials.

Generally speaking, the longer a country's television industry has aired commercials, the shorter the length of the advertisement has become. For instance, in countries where advertising is a new phenomenon, each individual commercial, on average, will run at least one minute in length and often more, due to the lack of sophistication of the advertising industry to shape the messages, the lack of the production industry/television outlets to produce the commercials, and the lack of television-advertising-sophistication of the viewing audience. It simply takes time for industries unaccustomed to crafting television advertising messages, and viewers unaccustomed to receiving those messages, to be able to connect in the short period of time required of a television commercial.

However, as the sophistication of all three groups grows, the length of the commercial is reduced, generally, to 30 seconds. It takes a certain amount of advertising sophistication to be able to craft a coherent commercial message into only 30 seconds and make it understandable to an audience. Further, the audience must have a reasonable amount of experience with comprehending television commercial messages to be able to add the nuanced portions of the commercial that have been left out to reduce the time to 30 seconds. While in the early days of television, the time necessary for the development, implementation, and comprehension by the television audience of 30 second commercials was many years, given today's technological advances and the ability to view all manner of micro-videos, the length of time from the beginning of commercials on television to the use of the 30 second commercial can be mere months.[3]

Finally, for those countries with highly sophisticated advertising industries as well as highly sophisticated viewers, and with a long history of advertising, television commercials can run for as short a period as 15-20 seconds. This ability to run such short commercials is due to the sophistication of the viewer who has watched literally millions of commercials over his/her lifetime and whose understanding of what a television commercial is makes it possible for the viewer's mind to "fill in the blanks" in a 15 second commercial. There are very few countries of the world where 15-20 second commercials are successful or even attempted, with the United States being the standard bearer.

Additionally, commercials can be placed within programs or between programs, although the latter is becoming less frequent as television stations and cable channels attempt to keep their audiences "flowing through" the viewing day (and especially the prime-time hours of 8:00 p.m. – 11:00 p.m. ET). Today, the legacy media outlets run many, if not most, of their programming right to the beginning of the next program to keep the audience from changing channels or moving to over-the-top (OTT) offerings such as Netflix, Hulu/Hulu+, etc. Further, throughout the history of television advertising, it has been the practice of programmers to make sure the same commercial goes out to the entire audience, regardless of the makeup of the audience. That remains the primary delivery method even today for legacy media outlets, although there are occasional instances where advertisers are willing to find ways to segment the audience and send a different commercial to each segment. Legacy media most often attempt this type of segmented advertising through the OTT offerings. The outlets that do are the vanguard of 21st Century Television advertising.

Let's take a look at how television advertising developed – first in the U.S. and then in other countries as well.

History of United States Television Advertising

Television and television advertising have been together since 1939 when television was first introduced to the American audience at the 1939 New York World's Fair.[4] Television advertising grew out of the radio advertising of the 1930s. So U.S. television, like its predecessor radio, was unique in that it was advertiser-driven from the beginning of its existence.

The first official, paid television advertisement (commercial) was run in the United States on July 1, 1941, by the Bulova Watch Company. It was a very simple, understated, and popular ad running all of 10 seconds. The ad showed the face of a Bulova clock superimposed over a map of the U.S. with a voice-over saying "America runs on Bulova time." The commercial was run on the New York station WNBT (later WNBC), and was run just before a major league baseball game between the Brooklyn Dodgers and the Philadelphia Phillies.[5]

However, advertising on television was short-lived. Once the United States entered World War II, television basically went into hibernation until the war ended. Very few stations remained on the air for any length of time during the day, and many television stations were used as facilities for teaching groups how to conduct civil defense drills.[6]

After the war, all aspects of the television industry began to grow rapidly as the U.S. moved from a wartime economy to a peacetime economy. Television advertising grew rapidly, as did the number of TV sets and the size of TV audiences. As an example, in 1948 less than one percent of U.S. homes had television sets; by 1952, one out of every three homes had TVs. Likewise, in 1949, spending on television advertising was $12.3 million; two years later in 1951, TV ad spend had grown to $128 million, and by 1954, television had become the leading medium for advertising.[7]

Sponsored Programs

Because TV advertising grew out of radio advertising, television programs initially had only one sponsor who received all the ad time during the show. Often the title of the program included the name of the sponsor. For instance, some of the top programs during that early era included the *Goodyear TV Playhouse; The Colgate Comedy Hour; Kraft Television Theater;* and the *Texaco Star Theater,* whose name even described the logo of Texaco (the Texaco star).[8] Even newscasts were sponsored. One of the best-known and most famous, ran on NBC beginning February 16, 1949. The *Camel News Caravan* was a 15 minute newscast that ran Monday through Friday from 7:45-8:00 p.m. Eastern Time throughout the East and Midwest. John Cameron Swayze – later of Timex Watch spokesman fame – was the news anchor. The program was sponsored by R. J. Reynolds Tobacco Company, maker of Camel cigarettes,[9] and the contract called for there to be a burning cigarette on camera the entire time.[10]

Virtually every program on TV had a sponsor. The sponsor got all the commercial minutes, the actors/actresses used the sponsor's product, and the sponsor had final approval of any creative decisions.

Spokespersons, Demonstrations, and Animation

Spokespersons were extremely important in the early days of TV advertising. The spokesperson could be seen as well as heard, and a good spokesperson could develop a relationship with an early TV audience. The prototype of the early spokesperson was Betty Furniss. Ms. Furniss, an actress of very modest credentials, became everyone's favorite spokeswomen during her time advertising Westinghouse products.[11]

Demonstration ads were extremely popular because people could now see the product being used. It was a powerful persuasive technique, especially to early TV audiences. The spokesperson would show the product in use, demonstrate how to use the product, and show the results of the product, depending on what type of product was being demonstrated. For example, Furniss would demonstrate a Westinghouse refrigerator by opening each compartment and explaining all the advantages of that compartment. Her demonstrations were both informative and exciting to early television watchers, especially housewives.

In addition to spokespersons and demonstration advertisements, early forms of animation were extremely popular. One of the earliest was Alka-Seltzer, whose "Speedy Alka-Seltzer" cartoon character, who looked very similar to a puppet but with no strings, had an Alka-Seltzer tablet for a body (with the product name on it) and a second one for a hat.[12] Compared to today's advertising the earliest ads were very simplistic. Additionally, well-known television cartoon characters such as Fred Flintstone and Barney Rubble were used as spokespersons for cigarette ads (Winston) in the early days.[13]

The 1950s sponsorship of television programs culminated in the "Quiz Show" scandal of 1957. Quiz shows had become extremely popular and they were all sponsored. Two of the more popular were *The $64,000 Question* – which moved from radio to television and increased the prize money from $64 to $64,000 - and *21*. It was the show *21* that was the focal point of the quiz show scandal.[14] In the Congressional hearings the sponsors blamed the networks and the networks blamed the sponsors. The only real losers were the champions, the hosts, and the producers. As a result of the scandal, Congress made it illegal to rig a quiz show without telling the audience it was rigged.[15] That's one reason for all the disclaimers at the end of each quiz show today.

The quiz show scandal forced the networks to quit having single sponsors for programs and brought about the type of advertising that is seen on TV today – the "magazine concept." The magazine concept, first introduced by Sylvester "Pat" Weaver, president of NBC, was taken from the way advertising was done in magazines – the magazine sells ads to whomever will buy them, and sells full page, half-page, quarter-page ads. Likewise, the networks would sell to whomever would buy and they sold 60 sec. and 30 sec. advertising time slots.[16]

It was during the 1960s that television advertising re-invented itself, moving away from the sponsorships that had so damaged the reputation of the networks, to controlling their content themselves and implementing Pat Weaver's "magazine concept" of advertising. The economy was expanding during the 1960s and so were the revenues of the television industry. From 1959 to 1969, television advertising revenues increased from $1.5 billion to $3.5 billion, a 230+% increase during the decade.[17]

It was also during the 1960s that color television became the television in most people's homes. Color television gave a more lifelike feel to programming as well as to television advertisements, leading television advertisers to embrace color commercials even more quickly than their customers. However, due to the increased production costs associated with producing color television advertisements as well as increased costs of the ad time itself, advertisers moved to developing the more-compact 30-second commercial.[18]

Things began to change in the 1970s, but television advertising continued strong. In 1971, Congress outlawed television advertising of cigarettes, a very lucrative revenue stream for the networks which ran about $150 million in 1970 alone. Television was quick to cover their losses from cigarette advertising by increasing their prices for 30-second spots to make up for the shortfall.[19]

For the first half of the decade, the television networks "owned" television viewing. Their competition was simply each other. Public television was transitioning from the educational television it had been in the 1950s and 1960s to the alternative it is today, but that growth was very slow and uneven. Cable, which had begun as community antenna television during the 1948-1952 television freeze as a way of providing local and distant television stations to communities in the "TV-freeze gaps," was still a minor competitor to the three powerhouse networks and their programming.[20] Additionally, completely independent stations dotted the television landscape, but they, too, were no match for the networks and their affiliates. For television in the first half of the 1970s, it was a sellers' market. Ratings points were critical for the networks, because during this time, each additional ratings point equaled an extra $75 million for a network during prime time.[21]

That all changed during the second half of the 1970s, beginning with the development of satellite cable delivery possibilities. HBO, Turner's superstation WTCG-TV (now TBS), and Pat Robertson's Christian Broadcasting Network (CBN) pioneered the use of geosynchronous satellites to carry their signals to cable systems all across the country. The development of satellite delivery of cable ushered in a "cable gold rush" era of unprecedented growth and the development of the cable multichannel universe seen today. The development also has led to the ability of advertisers to get their messages to much more specific audiences as the number of niche channels became a reality.[22]

Today, of course, television advertisers have myriad choices of where to place their advertisements, from legacy media to the latest in new media technologies. No longer will the networks and cable channels be able to dictate to the advertisers; going forward, it will be the viewing public that will dictate where advertisers should place their advertising as will be demonstrated in the coming chapters.

As mentioned at the start of this chapter, traditionally, the commercial has followed a pattern:

1. to start, commercials that are one minute in length;
2. as sophistication with commercials grow, they drop to 30 seconds

3. for highly sophisticated viewers, commercials can be 15-20 seconds

The sophistication of the viewer and the more experienced the viewer is with ads has determined the sophistication of the ad and the length of the ad.

As an additional reminder, commercials can be placed within programs or between programs. Further, the same commercial goes out to the entire audience, regardless of the makeup of the audience. Because of this "shotgun" approach to commercials, the average commercial has about a 15-20% success rate.[23] Advertisers pay huge sums of money to reach one in five viewers at the most.[24]

TV advertising is still the most powerful advertising tool today of the legacy media. However, online advertising continues to grow and estimations for 2018 called for total digital ad spend to far exceed TV ad spend, $107.30 billion (digital) to $69.52 billion (TV).[25]

Enter 21ˢᵗ Century Television advertising, the way advertising will be handled in the coming years. It will be different, exciting, and extremely profitable for everyone. Twenty-first Century Television advertising has three major requirements:

1. It must be targeted to the viewer
2. It must be interactive
3. It must be for the products the viewer wants delivered in the way/ways (s)he wants it.

We'll discuss each of these requirements in later chapters.

CHAPTER 2

THE TELEVISION PLAYERS

Twenty-first Century Television is becoming and will be nothing like 20th Century Television. Mergers, acquisitions, new technologies, changing viewing habits – all these and more are contributing to what amounts to a revolution in the television industries. Further, new players are entering (or attempting to enter) the legacy media industries, while the legacy players are moving rapidly to compete with their online competitors. The result will be – for the next ten years at least – a continuous upheaval and redirection of what television is now and what television will be in the 21st Century.

While the traditional shotgun approach has taken the legacy television industry to its current multibillion dollar status, that form of advertising often does not work for the new, online competitors. Further, as will be discussed in the next chapter, the television viewer is changing as well. No longer is the television viewer simply willing to watch programs through appointment viewing and on specified schedules implemented by the legacy television players. Rather, today's viewers – and even more, the young viewers of today who will be the dominant viewers of the future – are demanding television and – when willing to accept – television advertising, on their terms. The end result will be a 21st Century Television universe of anytime, anywhere, on any device television.

This chapter will look at a variety of different television industries, both legacy and new television industries, and how they are changing, adapting, or not adapting to the coming 21st Century Television universe.

The Broadcasters

In the U.S., broadcast television is an industry that is privately (not government) owned and operated, driven by the need for revenue generated primarily through advertising. Since its beginnings in the late 1940s, the broadcast networks and their affiliate local stations have been the dominant provider of traditional television to the U.S. audience.

Never in the history of television has there been such an upheaval in the broadcast portion of the television industry. The three (now four, five, or six, depending on how you consider Fox Television, CW Television, and Univision) networks and their owned-and-operated and affiliated local stations across the country were, for those who grew up with the enjoyment of watching them, a stable presence in audiences' lives. The networks brought the viewer the day's news and kept her/him entertained with a variety of scripted programs, while the local stations delivered the network programs to their audiences and provided the viewers with the day's local news, sports, and weather, along with special programming for kids during weekday afternoons and Saturday morning. Before the mid-1970s, the networks and local affiliates had the viewing audiences all to themselves. If viewers were going to watch, they were going to watch one of the three networks on the local market affiliates or the networks' owned-and-operated stations. The networks and their local stations had to contend with public television stations and a smattering of local independent stations, if one or more was actually in the market, but neither of the two competitors was any threat, generally speaking.

Before 1975 and the advent of satellite-delivered cable television,[1] the three networks[2] garnered approximately 98% of the daily prime-time television viewing audience.[3] The networks competed only among themselves having nothing to fear from the public television stations and the smattering of independent stations that existed in primarily large cities. Public television was still trying to decide what it would become, providing a mixture of educational programs aimed at children to supplement[4] the school classes they were taking, "Julia Chiles" British programs,[5] and *Mister Roger's Neighborhood*–like programs. However, all that has now changed.

Today, broadcasters are facing a fight for their lives. Their entire relevance – especially among the age groups younger than the aging baby-boom population – is being brought into question. The audiences for the networks and their affiliates continue to erode as viewers continue to go elsewhere. Where elsewhere? Well, everywhere elsewhere. Whether it is the myriad of narrowly-focused channel options on cable or direct-to-home satellite services, or through other alternative delivery systems ("alternative" to over-the-air), almost nine out of ten viewers get their television by some means other than the traditional over-the-air television.[6] With cable, DTH satellite delivery, over-the-top (OTT) set-top boxes, Blu-ray and DVD players, digital video recorders such as TiVo, as well as the online options such as Hulu/Hulu+, Netflix, Amazon Prime Instant Video, and YouTube, viewers are going anywhere and everywhere to "watch TV."

Beginning in 1975 with the satellite delivery of the *Thrilla In Manilla* championship boxing match, and continuing throughout the late '70s and '80s, cable developed into a "heavyweight" contender for audiences as the cable became populated with niche channels designed to take specific groups of viewers away from the broadcasters. Now the audience had more and more choices of programs to watch, and the broadcasters saw the size of their audiences shrink dramatically. However, because each of the niche channels had tiny (but often very loyal) audiences, it was hard for advertisers to justify moving large portions of their advertising budgets from broadcast television to cable television. Nevertheless, for the broadcasters, those myriad cable channels – because there were so many of them – continually drew portions of the audience away from the broadcaster, leaving them with smaller and smaller audiences for their programs each year. Even the most popular of the broadcast programs today – network or local station – have audiences so small compared with the programs in the pre-1974 time that most, if not all, would have been cancelled instead of lauded as they are today.

At this time, the networks are fighting for their existence. Every year, the networks – whether it's the traditional Big 3, or the Big 3 plus Fox, CW, and Univision – continue to lose market share as more and more viewing alternatives become available to the television consumer. Today the networks have to contend with numerous cable or direct-to-home satellite channels, many telecasting reruns of network programming, while others provide original programming that have special interest to niche, but important audiences. Additionally, the networks must contend with those new media that deliver their programming through the Internet to audiences' homes not only to their computers, but on the go to their tablets, laptop computers, and mobile phones. Further, the networks have to contend with viewers' big-screen home television sets that connect directly to the Internet, or that can connect through one of the many different types of over-the-top set top boxes.

Whether it is Netflix or Amazon Instant Video offering movies, television programs, and original programming streamed to the television on demand through a variety of devices, iTunes providing downloads of movies and television programs, user-generated programming through YouTube being delivered, or even their own programs being streamed through Hulu/Hulu+, the networks and their local affiliates are having to battle competitors, many of which were not around as recently as a decade ago. Further, the cable and DTH satellite companies themselves have recently begun to offer movies and programs as video-on-demand for a fee, uncut and uninterrupted, earlier even than HBO and the other premium pay-television channels. Finally, the networks and their local affiliates must compete with original programming from not only HBO, Showtime, and the other premium pay-television cable offerings, but also from Internet sites such as Netflix, Amazon Instant Video, Hulu/Hulu+, as well as Yahoo and YouTube, which have begun specialized channels in their Internet space for long-format, professionally-produced programming.[7]

At this time, the networks' local over-the-air affiliates are embroiled in a battle with the mobile telephone industry and the Federal Communications Commission over bandwidth usage by the affiliates.[8] Traditionally, every over-the-air television channel was provided six megahertz of bandwidth on which to deliver its signal to the home. Today, with the various means of compressing that same television signal without loss or degradation of signal quality, television stations need only a portion of the original six megahertz of bandwidth to broadcast their signals. With digital television – as television is today – the programmer can adjust the amount of space needed for a program, making it possible for several compressed channels to be located within one analog channel and still look high quality to the viewing public.[9]

Because of the proliferation of smartphones and tablets capable of surfing the World Wide Web; sending and receiving e-mails, texts, and multimedia messages; and downloading and playing music, movies, and even television programs in their entirety, the amount of bandwidth set aside for the mobile telephone industry is dwindling rapidly. The largest amount of additional bandwidth is the extra bandwidth the television stations have as part of their original six megahertz. However, because the broadcasters are in a furious battle for survival against **all** their competitors – including the mobile phone and tablet computer industries – the broadcasters' willingness to give up that extra bandwidth, even if they are paid for it, is extremely mixed. Regardless of whether or not individual stations are willing to give up part or all of their bandwidth, overall the broadcast industry simply is not ready to give in to the inevitable – the day of over-the-air broadcasting is dying and will not return.

This refusal can be seen in the broadcasters' latest attempt to control their future and the future of their competitors with the introduction of ATSC 3.0, a set of standards that, if implemented, will allow the broadcasters to deliver high speed broadband to homes through the airwaves on the unused portions of their channels' bandwidths. Further, because the broadcasters are advertiser-driven, discussions are on-going that may make ATSC 3.0 available to consumers for little or no cost whatsoever, thus undercutting significantly their home broadband competitors.[10] Should such occur, the broadcasters would have a virtual monopoly on the delivery of the Internet – and any and all forms of Internet TV/video to consumers' homes. Broadcasters could then decide which of their content provider competitors to allow to continue and which ones they would block or price out of business, likely making Netflix, YouTube, etc., no longer available except by Wi-Fi or cellular service to smartphones (primarily), potentially setting back the development of 21st Century Television by decades.[11] Further, because ATSC 3.0 would be delivered by the local stations in each market, their ability to block the streaming services of their networks would also be an option, moving the local stations back into the power positions they were in during the 1960s and 1970s, and seeing the networks held hostage to their affiliates and being forced to surrender their streaming options as well.

So where does the broadcast industry go from here? What is its place in the 21st Century Television universe? Much depends on whether and how ATSC 3.0 develops. Should ATSC 3.0 become the delivery mechanism of choice for the U.S. population at large, the broadcast industry will be able to control its own destiny and name its own price to all other television technologies and content providers. If ATSC 3.0 ends up going the way of Mobile DTV, the broadcasters' initial attempt at competing in the streaming arena (an abject, and for some, expensive, failure), then they will become another player – albeit, a powerful one should they so choose – in the 21ˢᵗ Century Television universe of content providers.[12]

Networks and Local Stations

Now let's move away from ATSC 3.0 and look at where the networks and local stations are heading otherwise. Without the development of ATSC 3.0, the networks and their local affiliates will be forced to move in extremely different directions. For the networks, they will most likely move toward developing their streaming services, both SVOD[13] and AVOD[14] to better compete with the numerous other competitors for space in the 21ˢᵗ Century Television Universe. They will be forced to for the following four reasons:

1. The current delivery system is no longer cost-effective, especially in the U.S.
2. Technology makes traditional delivery of broadcast television obsolete.
3. Providing direct delivery of programming to consumers is more profitable and opens more avenues for revenues.
4. The up-and-coming generations do not watch broadcast (over-the-air) television. Increasingly, they don't watch any of the legacy media, preferring true 21ˢᵗ Century Television platforms.

Looking at the four reasons mentioned above, it is easy to see that the broadcast industry must begin to, or continue to, make major changes in the ways they do business if they wish to be competitive and remain the major force in the industry. As such, the networks and the local affiliates, ultimately, will go their separate ways. The networks had begun the process with Hulu/Hulu+, owned by Fox (30%), ABC/Disney (30%), and Comcast/NBCUniversal (30%), along with a small stake owned by AT&T (10%). That joint ownership no longer exists as of May 2019, when Disney finished buying 100% of Hulu/Hulu+ – the last purchase coming from Comcast.[15] CBS did not participate in the joint ownership of Hulu/Hulu+, preferring to go alone with its CBS All Access streaming service.[16] In addition, each of the networks have their own individual apps and the other networks have developed, or are developing, their own streaming services, in the realization that streaming is the future of television..

Of these beginnings, CBS All Access is the prototype of the new 21st Century Television network. Even so, this subscription-only service that currently serves as an ancillary part of CBS' traditional broadcast network will become the major delivery system for CBS, with the future being not only its current SVOD service, but also an AVOD service as well. The AVOD service will likely supplant the SVOD service as the service of choice for viewers of 21st Century Television.

For the local network affiliate stations, the future is bleak – there's no other way to say it. In the U.S. – unlike most other countries around the world – television has been delivered to its audiences through the use of a highly developed and vibrant local TV station industry. The television model in the U.S. grew out of an excellent radio industry with its networks and local affiliate stations going back to the 1920s.[17] The country was divided into non-overlapping (at that time, for the abilities of the equipment) geographical areas commonly known as "markets," with local affiliates of all three networks[18] eventually filling the needs of the viewers in each market. Since 1952, the local affiliates have been delivering the networks' programming as well as their own local programming – basically local newscasts – to their loyal, and captive, audiences. That is changing dramatically.

With the success of SVOD services like Netflix, Hulu/Hulu+, Amazon Instant Video, and others, and with the developing successes of CBS All Access, HBONow, as well other current, new, and developing direct-to-consumer, IPTV-driven[19] choices, the local stations are becoming, and very shortly will be, superfluous to the networks they have so well-served for so long. They are the major casualties of 21ˢᵗ Century Television. It is the local stations that lose when each of the networks drop them and go straight to the consumer.

When (no longer "if" as has been described in so many articles and even in previous books by this author) the networks drop their affiliates, the local stations have three choices:

1. Become the video portion of a local newspaper
2. Become the video portion of a cable company
3. Become a hyperlocal independent TV station

Of those three choices, the third choice – become a hyperlocal independent TV station – will be the fate of the vast majority of the local former network affiliates. Those that survive will operate direct-to-consumer over IPTV and will become, in a sense, the television version of today's local radio station – highly local and focused solely on the unique needs of the market it serves. The local station industry will ultimately shrink by 2/3 to 3/4 with the remaining stations mostly becoming hyperlocal stations.

The Legacy Pay-TV Industry

The legacy pay-TV industry today consists of the cable industry and the direct-to-home satellite industry. Both of these industries can be considered to be in free-fall with regard to their subscribers, of which the cable/DTH satellite delivery portions is, and will be, most affected by the 21ˢᵗ Century Television revolution. Additionally, though, the cable/DTH satellite content provider portion will be affected by this same revolution because these content providers must compete directly with each other, the broadcasters, and the new television content providers (such as Netflix, etc.) for viewers.

Although the cable industry began during the television freeze of 1948-1952 as community antenna television,[20] it wasn't until HBO telecast the heavyweight boxing match from the Philippines using satellite delivery to bring the fight to the American public that cable moved from a retransmitter of broadcast television to a true competitor for the broadcasters. At its height in 2005, cable penetration in the U.S. was just slightly less than 71%.[21] Today, as of the 3rd quarter of 2018, cable penetration stands at approximately 40%, a roughly 43+% reduction in just 13 years.[22] However, even with the reductions, today, in the U.S., the cable industry has the largest percentage of households of any form of television.

The cable industry today is undergoing tremendous, dynamic change. Cable in the coming 21st Century Television universe will little resemble the way it looks today. In fact, the cable industry has been one of the most forward-looking of all the legacy industries, because the industry realizes that it must change to meet the new and developing technologies capable of delivering television to the consumer. The cable industry will become, if not irrelevant, then certainly marginalized if it does not embrace the 21st Century Television universe.

Today, of course, cable delivers to its subscribers hundreds of channels, both over-the-air and strictly cable channels (not over-the-air). There are channels for just about everybody, including music channels and video-on-demand channels, in addition to news, sports, weather, and entertainment channels. In addition, cable provides subscribers premium services such as HBO, Showtime, etc., for additional costs. Each of the premium services has a number of channels, often aimed at specific audiences. Generally, the more premium channels a subscriber purchases, the less the cost for each channel.

Regardless of whether in the U.S. or elsewhere, cable today faces life-changing competition from some, or all, of a variety of different sources. These include:

1. Direct-to-Home Satellite services
2. Internet Protocol Television
3. Streaming of legacy media programming using Hulu/Hulu+ and/or apps
4. Streaming of Internet-only services such as Netflix, Amazon, YouTube, etc.
5. Over-the Top Set Top Boxes & Connected "Smart TVs"
6. Cellular Television

Let's look at these separately:

1. DTH Satellite – Both DirecTV and Dish Network are legacy players like cable and the broadcasters, but what once was a complement to the cable industry for people living outside the coverage areas of cable, has now become its most powerful challenger. Before the development of low-earth-orbit (leo) satellites, DTH satellite services generally claimed about 5-10% of the pay-TV audience. The dishes were bulky and expensive, and they created an unsightly situation for those who had to place the dishes in their front yards. However, with the development of the leo satellites, came DirecTV and Dish Network, the two major players in the DTH satellite space. Together, at their height in 2011, they had more than 30% of the viewing audience, and were the first to take subscriber share away from the cable industry. The share for the two services, though, is quickly dwindling.[23]
2. Internet Protocol Television (IPTV) - Internet protocol television, or IPTV, will be the delivery mechanism for all of

21st Century Television. The ubiquitous nature of the Internet, coupled with its versatility, makes the Internet the perfect powerhouse delivery system for television in the future. There is no doubt that IPTV and 21st Century Television go hand-in-hand. The move to an all-IP infrastructure has been, seemingly, on almost everyone's lips at the National Association of Broadcasters conventions since at least 2015. It is the driving force behind the broadcasters' attempts to develop ATSC 3.0 to compete primarily with the cable companies move from traditional cable to an all-broadband industry. When you have the **broadcasters(!)** constantly talking about the need to go to an all-IP infrastructure, is there really any doubt that IPTV is the future of the television universe? The versatility of IPTV makes it an excellent delivery system for television of the future. IPTV can be delivered over a wired connection, through Wi-Fi, or through a cellular connection. It can deliver high, ultra high (4K), and super high (8K) definition television, 3-D television, and holographic 3-D ultra high definition television, either through a traditional scheduled day of programming, or in an on-demand format. IPTV can deliver completely addressable, interactive advertising through the use of aggregated targeted microadvertising, programmatic media-buying, and dynamic ad insertion.[24] It can deliver a wide range of differentiated and undifferentiated promotional content designed to attract the viewer to those television programs that (s)he wants to watch as well as the products and services the viewer/consumer desires.[25] Today, IPTV is a competitor to cable; tomorrow it will be the underpinning technology for delivering 21st Century Television.

3. Streaming of legacy media programming – Today, of course, streaming legacy programming competes with

cable in that, with Hulu/Hulu+ and apps such as CBS All Access, HBOGo/Now, ESPN+, etc., viewers no longer need the cable service to watch their favorite legacy channel programs. By subscribing to just a few of the services available to viewers over the Internet, the consumer can watch her/his favorite programs at a fraction of the cost of what a full cable or DTH satellite package would cost each month. Further, with the proliferation of the "skinny bundle,"[26] the ability to enjoy all the content providers the viewer desires becomes easy and more cost-effective.

4. Streaming of Internet-only services – While there are myriad choices for services that are available across the Internet, the best-known and most-subscribed-to services are Netflix, Hulu/Hulu+, Amazon Prime Video, and YouTube. While each of these is a competitor to the cable services and to the cable industry in general, in truth, they add to the cable industry's revenues because they drive cable broadband subscriptions – the future of the cable industry. As the cable industry continues to evolve from the legacy industry it currently is to the broadband delivery system it will become, Netflix and the other Internet-only streaming services will encourage more and more viewers – especially younger Gen X, Millennial, and Gen Z viewers – to subscribe to cable broadband and Wi-Fi services to receive the programming on these SVODs.

5. Over-the-Top Set Top Boxes and Connected TVs – Both of these choices are the gateways between the cable systems and the Internet choices mentioned above. However, depending on the type of device it is, the set top box can also deliver gaming, play DVDs, stream music, connect to skinny bundles, surf the web, use social media, as well as connect to the Internet to deliver the choices mentioned in #3 and #4 above. Connected television sets serve as gateways

to deliver #3 and #4 above, and can also be connected to set top boxes to provide the ability to do the other choices just mentioned. They can also be connected to the cable or DTH satellite system to deliver those program choices. As such, the connected TV set is both a competitor and a partner with the cable industry.

6. Cellular TV – The idea of television being delivered through cellular towers not only to smartphones and tablets designed to receive the signals, but also to home television sets was first discussed in the 2012 edition of *21ˢᵗ Century Television: The Players, The Viewers, The Money*. It was discussed as a future technology and competitor to the cable and DTH satellite industries. In reality, little headway, if any, has been made to cellular television becoming a reality – at least publicly. That'll all change with the roll-out of 5G technology. At that point, everyone from the cable companies, to the mobile phone companies, to the major set top box companies, to Netflix, Hulu/Hulu+, Facebook, YouTube, and others, will be offering the viewers and/or subscribers the ability to receive television through 5G technology. Once 5G is fully deployed to an area, it will provide each of the mobile phone companies, such as AT&T, Verizon, and T-Mobile/Sprint, with a potential nationwide cable network without having to pull cables into the home.[27] The cable companies will certainly follow suit and the results will be astounding.

The future of cable is uncertain at this time. Cable is entering its declining years as a legacy media industry. For the cable industry, it's "change-or-die" time. Fortunately, the cable industry understands its situation and is moving to make the changes it needs to continue to be competitive in the future. In the short-to-medium term – the future of cable is not cable; it is broadband and Wi-Fi. Television is moving to an all-IP infrastructure, making IPTV the delivery system of television in the future. For cable to continue to be competitive during this period, cable must roll out a true "TV Everywhere" service that is not limited by geography or national borders. Today there are limitations on any cable company's TV Everywhere service. For cable to survive in some form, subscribers must be able to make use of their services anywhere they are. Notice the comment "not limited by geography or national borders." It is key for cable to move to a global mindset as it transitions to an all-IPTV industry. The future is global for everyone – cable included.

In the medium-to-long term, cable must become the delivery method for the coming Internet of Things – or more correctly "Internet of Everything." Smart TVs will give way to "smart homes" which will in turn, give way to "smart communities," and finally "smart cities." The cable industry can be the cornerstone of IoT or IoE, but only if it chooses. As such, it must also embrace the widespread use of 5G (and any further generational upgrades – 6G, 7G, etc.), and become a leader in that space as well.

Finally, in the long term, the cable industry must become a single world-wide industry, by combining with other cable and IPTV companies around the world. While the cable industry is a highly consolidated industry today, it must continue to consolidate further, joining together with other cable and mobile phone companies around the world into a global whole. Otherwise, if the industry continues on its "local monopoly" view of itself, it will be bypassed by the newer technologies industries and those current industries that are willing to innovate, until it no longer exists or it is a mere shadow if itself, trying to cling to whatever life it may have left before its flame ultimately dies away.

In summary, the cable industry is at a crossroads in the U.S. and, in reality, in countries around the world that have active cable industries. Either they move to IPTV and consolidate globally, or they will become increasingly irrelevant.

Direct-to-Home Satellite Industry

The Direct-to-Home (DTH) satellite industry is one that has been much more important around the world than in the U.S., given the influence of the U.S.'s cable industry. In many parts of the world, DTH satellite systems have been the preferred, or even only, way for viewers to watch something other than over-the-air programming. Today, slightly less than 3 out of 10 television viewers in the U.S. watch their programs using one of the two companies to deliver the programming. That percentage had been stable for a number of years, but is now seeing even steeper declines in subscribers than the cable industry.[28] Even so, for the time being at least, DTH satellite is the major competitor to cable. That is bound to change in the next very few years.

Worldwide, virtually all the countries of Europe, North America, most of the countries of South America, and Australia and New Zealand use DTH satellite for their major delivery alternative to OTA television. Further, there are 17 countries in Asia with DTH satellite delivery. In fact, the only continent lacking in direct-to-home satellite service is Africa, where almost no countries have DTH systems. Interestingly, Europe led the way in developing the DTH satellite industry, with Luxembourg offering the first service in December 1988.[29]

Direct-to-home satellite systems suffer from three major problems:

1. The inability to easily offer landline or mobile phone service;
2. Latency (time lag) problems, which makes it virtually impossible to have truly fast, high-speed/high quality broadband service; and
3. Weather issues – when there are thick clouds between the satellite and the receiving dish, there can be no signal at all until the clouds pass.

The future of the DTH satellite industry as a stand-alone industry is bleak. As the world transitions to Internet Protocol Television, the DTH industry will suffer. Ultimately, the different DTH satellite companies will be forced to merge with other companies that offer complementary services in order to survive. Even that may not be enough to save the DTH satellite industry. DirecTV has been bought by AT&T to provide subscribers with satellite television because AT&T is having trouble rolling out its U-verse service nationwide. However, AT&T has introduced its DirecTV Now skinny bundle, and is advertising it against its own traditional DirecTV offering. Dish Network is looking more toward 5G to bolster its business future.[30] Regardless, with more and more skinny bundles, OTT offerings, alternative choices, and cord-cutters/cord-shavers/cord-nevers, the likelihood of the DTH satellite industry continuing to be a force in 21ˢᵗ Century Television is slim to none.

The OTT Alternatives

Over-the-top (OTT) delivery of television is destined to be the future of 21ˢᵗ Century Television. As consumers – and especially the younger consumers – want their television anytime, anywhere, and on any (every) platform, legacy television will have to change or become irrelevant and even cease to exist. Whether it's the powerhouse OTT content providers such as Netflix, Hulu/Hulu+, Amazon Prime Video, YouTube in its many paid and free choices; the major legacy providers such as CBS All Access, HBO Now, ESPN+, or Showtime Anytime; or the myriad niche OTT channels that reside throughout the Web, viewers have an almost unlimited number of choices to hold their attention spans and babysit their children. If a person's preference is not a la carte, then there are the OTT content distributors delivering skinny bundles that can provide customized packages of channels designed just for that person. Choices such as Dish Network's Sling TV, AT&T's DirecTV Now, Sony's Vue, Amazon's Channels, Vimeo, as well as numerous others large and small are available to the viewer who is not quite ready to pay for each channel individually, or who likes the idea of a package as long as it's to her/his liking.

There are also the hardware manufacturers that provide the choices of how the consumer will receive his/her content. Whether it's an OTT set top box from a company such as Roku, Apple, or Google; it's a connected (smart) television set from LG, Samsung, Vizio, or others; smartphones, tablets, gaming systems, or VR/AR headsets – consumers will have plenty of options when viewing/experiencing their favorite 21st Century Television programs. And all will be driven for the foreseeable future by the force known as Internet protocol television, or IPTV.

Internet-Protocol Television will be the future of 21st Century Television. The ubiquitous nature of the Internet, coupled with its versatility, makes the Internet the perfect powerhouse delivery system for television in the future. IPTV can be defined as "multimedia services such as television/video/audio/text/graphics/data delivered over IP-based networks managed to provide the required level of QoS/QoE, security, interactivity, and reliability."[31]

IPTV is the underpinning for the coming new golden age of television because IPTV offers a number of advantages that other ways of delivering television simply do not have. Let's look at each of these advantages in some depth:

1. **The potential for total interactivity within the television landscape** – IPTV is interactive TV. Since the earliest days of television in the 1950s, there has been a desire to have the viewer be able to interact directly with what (s)he is watching. This desire for full interactivity becomes a reality and not a dream because of the capabilities of IPTV. The ability of full interactivity makes it possible for advertisers to engage audiences in new and special ways, and programmers to get instant feedback on their choices of shows.

2. **Television viewing from a "lean back viewing" to an immersive, completely interactive "lean forward experience."** Traditional viewing of television is in the "lean back" mode, where the viewer sits back and lets the program wash over her/him. Television is in the earliest days of moving away from the traditional way of watching television to a more immersive "lean forward" way of experiencing television. Notice the difference in the two terms: "lean back" is watching; "lean forward" is experiencing. IPTV can bring the lean forward experience to the audience member (no longer "viewer") today through the early use of virtual reality (VR) headsets and tactile equipment, and, in the future, to holographic television.[32] With IPTV delivering the television programming, the audience member will have the choice of how to enjoy television – watching, experiencing, or somewhere in-between.

3. **Ability to deliver all television choices to a variety of devices and screens.** Because IPTV is Internet-driven TV, audience members will be able to enjoy their programs their ways. IPTV can deliver television to the traditional television set through over-the-top set top boxes such as Roku, Apple TV, Chromecast, and Amazon Fire TV, a video game console such as an Xbox or Playstation, a Blu-ray DVD player, or directly, when the television set is a connected, or "smart" television set. It can deliver a television signal to the traditional personal computer through wired, wireless, or cellular connections. IPTV also has the capability to deliver television to mobile devices such as smartphones and tablets, and notebook computers so that the viewer-on-the-go can enjoy his/her favorite television program.

4. **Ability to multitask while watching television right on the television screen.** As an audience member enjoys a favored program (s)he can surf the web; check e-mail; connect with

friends, neighbors, family, and other viewers; find out more about the star of the program or a supporting cast member; purchase a product seen on television in an advertisement or through product placement; search for other programs the viewer might enjoy; play a video game – all simultaneously. In other words, IPTV allows the audience member to do most anything (s)he would like without having to use multiple devices or stop doing one thing to do another.

5. **Ability to watch any television program produced by any television company and shown on any content provider anywhere in the world at any time and on any device.** IPTV gives the viewer the opportunity to follow his/her favorite program no matter where in the world the viewer is, no matter where the program originates or is distributed through. For instance, an American viewer, traveling in Santiago, Chile, could watch a program that originated in the U.K. and that's telecast on the BBC on his/her tablet or computer. That same BBC program could also be watched by a Vietnamese at home on his connected television set, while his neighbor, who happens to be Thai, is watching that same program on his/her smartphone while traveling in South Africa. All three viewers are sharing a common experience. While each is a native of a different country, they can enjoy the same program on different television technologies, in different locations around the world at the same time. Only with IPTV is that possible.

6. **Ability to drive advertising in ways that are only being explored today.** The ability to deliver addressable advertising – advertising that is specific to the viewer – is possible through the use of IPTV. It makes it possible to deliver one advertisement to one viewer who is watching a television program, while delivering a second advertisement to a second viewer who is watching the same program at the same

time, and a third advertisement to a third viewer watching the same program at the same time. Each advertisement can be tailored to the viewer's likes and dislikes so that the ad has a virtual certainty of making an influence on the viewer.

7. **Makes possible the development of future technologies that are on the horizon or even on the drawing boards of today's start-up companies as well as the major television players.** VR, cellular television, holographic television, television viewing on the windshields and windows of autonomous automobiles – these are just a small sample of the abilities of IPTV for the future. It will be exciting to see what will come next.

If IPTV will be the delivery mechanism of 21ˢᵗ Century Television, video-on-demand (VOD) will be the programming strategy, if you can call it a strategy. VOD makes it possible for TV Everywhere to exist. It frees the viewer from the tyranny of the television programmer.

From the beginning of television, the viewer has had only one choice for watching TV programs – being slaves to the television programmer. The programmer decided what the viewer would watch, when (s)he would watch, and, by choosing programs, on what channel the viewer would watch.

Video-on-demand changes the relationship between the programmer and the viewer, by giving the viewer the power to determine what, when, where, and on what platform (s)he chooses to enjoy television. It makes it possible for viewers to be able to watch their favorite programs anywhere in the world. The content can include user-generated content (YouTube), professionally produced programs on branded channels (YouTube and Yahoo), time-shifted over-the-air programs (network websites and Hulu/Hulu+), and movies and original programs seen on specific VOD services (Netflix).

VOD eliminates the need for the broadcast and cable/satellite channels (the legacy content providers) to cancel programs, because of too few viewers or the viewership is declining. Instead, programming departments will only have to be concerned with purchasing the best programs for the audiences they are trying to reach, because, once purchased, the programs will always be available. Additionally, with VOD, promotion departments become even more critical to the content provider because the only way viewers know about what's available is by continuous and ubiquitous promotion.

The move to VOD has already begun, with Netflix leading the way by serving 190 countries worldwide. In an effort to compete with Netflix directly, or to duplicate Netflix's success, more and more content providers are now making their programming available online through their websites and apps.

If IPTV is the delivery system for the coming 21ˢᵗ Century Television, and video-on-demand is the delivery strategy, then the OTT choices are the hardware that the viewer will use to enjoy his/her favorite choices. Whether it's through a stand-alone set top box, a videogame console, a Blu-ray DVD player, a connected television set, a smartphone, tablet, or computer, or a VR headset – and, later, holograms – the viewer has a wide array of choices to receive her/his television. Each of these devices makes it possible for the viewer to enjoy his/her favorite program, and, depending on the device, that enjoyment can be anywhere, anytime, and on the platform of choice or that's available. Further, each of these devices will be able to deliver fully interactive viewing, addressable advertising, ubiquitous product placement and global reach that is the essence of 21ˢᵗ Century Television.

Summary

21ˢᵗ Century Television will be global in nature, with every content provider becoming a global provider. Advertising will be generated from worldwide sources, leading to major new growth in revenues. Competition will lead to extreme worldwide consolidation of companies, leaving television in the hands of a small number of transnational corporations.

Nevertheless, the possibilities of 21ˢᵗ Century Television are huge and continually expanding. The developments in technology will make it possible for the television of today to seem as quaint and anachronistic as black-and-white television of the early days of television seem today.

Within the next few years, to the youngest of today's television viewers, the thought of "watching" television will be boring and old-fashioned. They will want to "experience" all that 21ˢᵗ Century Television will have to offer. It will be amazing!

CHAPTER 3

THE TELEVISION VIEWERS

The 21ST Century Television viewer can be defined as anyone who watches television and who has enough knowledge or willingness to learn about the upcoming television possibilities. In other words, almost anyone who is younger than about 80 these days. By 2020, that definition will cover almost everyone in the U.S. and the developed world countries.

These viewers will have a knowledge of how to use a computer; what the Internet and World Wide Web are; what a website is; what Google, Apple, Windows, Android (and others) are; have heard of – and most likely watch – YouTube or other user-generated-content (UGC) sites; have at least heard of, and most likely used – an Internet-streaming site such as Netflix; have heard of – and most likely have – a HD/4K/8K television set, a DVR, and likely a Blu-ray/4K/UHD player.

While most discussions of the viewing audience focus on all the different age groups from the "Greatest Generation" (those currently 73 and older, but mostly in their 80s or older), through the Baby-Boomers, Generation X, Millennials, to the youngest age group, Generation Z, for purposes of this book there will not be a discussion of the oldest generation and less attention paid to the baby boomers. Rather, this section will focus on the younger generations as they will be the ones most impacting and being impacted by 21st Century Television monetization.

As has been the case with previous books in this series,[1] this section will be primarily based on studies conducted by the Deloitte Corporation, a worldwide professional services firm based in New York City. The Deloitte Corporation is the largest private professional services network in the world. The company was founded in 1845 by William Deloitte in London making it also one of the oldest companies of its kind as well. Today, Deloitte operates in more than 150 countries worldwide, employing more than a quarter of a million people around the globe.[2]

More than a decade ago, the Deloitte Corporation began to look at how the media landscape was changing using their own resources to conduct a worldwide survey. That survey, originally called the *State of the Media Democracy Report* surveyed five different groups in both the U.S. and in representative countries around the world. Over the years the survey has evolved due to the changing media technologies that impact the viewing public. Today's *Digital media trends survey* is the 13th iteration of that original study. While the Deloitte studies have included five age groups – The Mature Generation, The Baby-Boom Generation, Generation X, The Millennial Generation, and Generation Z – because the monetization of 21st Century Television will have virtually no impact on the oldest of the age groups, this age group will not be discussed. Instead, this chapter will focus on the four remaining age groups (although Baby-Boomers to a lesser extent for reasons similar to the Mature Generation), and especially the three youngest – Generation X, Millennials, and Generation Z. Before looking at the current viewers of the generations, let's first understand the backgrounds of the Baby-Boomers, Gen Xers, Millennials, and Gen Zers.

Generational Backgrounds

The Baby-Boom Generation

What can you say about the Baby Boom generation? It is comprised of the largest number of children ever born in the United States until their children and grandchildren – the Millennials – came along. As such, the baby boomers have had a profound impact on all aspects of American society, the likes of which this country had not seen before and has not seen since.

The baby boom generation is the product of the wars fought during the middle decades of the 20ᵗʰ Century, specifically, World War II. Soldiers coming home to their wives and sweethearts wanted to continue or start their families and forget about the hell they had gone through. Their wives and sweethearts were more than obliging – thus, a generation was born that changed the landscape of the country with their births and continues to change society in dramatic ways. The general time frame runs from about 1946, the year after World War II ended, until about 1964 or 1965 – a span of about 18-19 years, depending on the source.[3]

For the baby boomers, society has been in a constant state of change. Their early days saw the likes of Dwight D. Eisenhower as president, Levittowns springing up across the landscape, and the early days of television. Boomers swelled the rolls of elementary and high schools, causing a boom time for book publishers and for contractors as they struggled to build enough school buildings to house all the youthful baby boomers.[4]

The baby boom generation has been the first generation to deal with massive changes in technology. When the first boomers were born, television was just coming out of a World War II experience where those few stations that had gone on the air before the war shut down for lack of parts, programming, and everything else. Television sets were black and white and housed in a piece of furniture the size of a small cabinet. The baby-boomers grew up with the early days of TV programming, watching the Saturday morning cartoons from the 1940s, complete with the fighting and violence those cartoons contained. In the evenings, the baby-boomers gathered with their families around the television set for an evening of television much as their parents – the mature generation – gathered with their families around the radio sets in the 1930s and 1940s to listen to many of the same programs in radio form that the baby-boom generation would watch on their televisions. Primetime programs included everything from *Perry Mason* to *The Untouchables* to Milton Berle's *Texaco Star Theatre* to the big name comedy shows that moved from radio to the new technology of television. By the late 1950s, baby-boomers were enjoying a "boom" in prime time Westerns. In 1958, the top four programs and five of the top six were Westerns, led by *Gunsmoke* and *Wagon Train*.[5] And then, of course, there was *I Love Lucy*.[6]

The baby-boom generation is the first true television generation, the first to embrace the new technology that was television. In the 1960s, the baby-boom generation saw their television viewing move from black and white to color and witnessed, on television, Presidential debates, the Kennedy assassination, and man's first steps on the moon – the culmination of a decade of space exploration. They also began seeing war footage from Vietnam, as many of the baby-boomers saw action in the jungles of Southeast Asia.[7]

During the 1950s, 1960s, and early 1970s, the three networks enjoyed the attention of more than 98% of the prime time viewing by the baby-boomers and their parents. The networks had only themselves to compete with, along with a handful of independent and public television stations that – in total – garnered maybe 2% or so of the prime time viewership.[8] Audience ratings and shares were high for the networks, with people watching the most television on Sunday nights and the audience percentage dropping throughout the week to its lowest point on Saturday nights. For most of the baby-boomers, there were three choices of programming at any given time on television – ABC's choice, CBS' choice, or NBC's choice. If you didn't like any of those three choices you either picked the least objectionable program or you did something other than watch television.

For the baby-boomers, all that changed in 1975, when HBO got the telecast rights to the "Thrilla in Manila" professional boxing match and had to determine a way to telecast the fight to the U.S. audience from the Philippines. HBO's satellite solution would revolutionize television forever and herald the beginning of a new television universe. The next year Ted Turner put his Atlanta-based UHF station on the satellite as well, and the cable television industry was off and running. With the oldest baby-boomers in their late 20s and the youngest just reaching ten years old, the baby-boom generation was the early adopter of cable, the first major change in television since the transition to color, and what would be the first of the major changes to television since it displaced radio as the major entertainment medium.[9]

The oldest baby-boomers are now in their retirement years (the oldest are a little more than 70 [71 specifically as of the current survey]) while the youngest are in their early 50s (53 specifically). They are world and country leaders, captains of industry, and the money behind the technological geniuses who are half their ages. They are the heads of the major corporations that run society throughout the world and are in charge of all things television.

Their children are the new media moguls, creating the next generation of television. Their grandchildren are the current consumers of the newest and most profound developments in all-things-TV. The baby-boom generation is the first true television generation, the first to embrace the new technology that was television. As such, most baby-boomers have had a lifelong love affair with television. baby-boomers have lived their lifetimes with televised war, beginning in Korea and going through Vietnam, to the civil wars and terrorism of today. Much as their parents before them, the Baby-Boom generation is a special generation, the entrepreneurs and drivers of 21ˢᵗ Century Television.

Generation X

The Generation X age group (1966-1982) is unusual in that it is the true transitional generation in these studies. Older Gen Xers are part of the "baby-bust" generation – the generation following the baby-boomers that saw a downturn in the number of children born in the United States. These baby-busters were born between 1966 and 1975. They were the children of an era when the birth rate in the U.S. took a nosedive, brought about – in part, at least - by the development of "The Pill," the Supreme Court's decision in *Roe v. Wade* - the court case that legalized abortions - and the move away from the traditional family lifestyle to a variety of alternative lifestyles, freed from sexual constraints by the free love generation of the 1960s. The result was the baby-bust generation, a generation where births in the U.S. dropped from 72.5 million during the postwar baby-boom years to 56.6 million, and where in 1975, the birthrate in the U.S. sank to 14.8 newborns per 1,000 citizens, the lowest birth rate in the history of this country.[10]

The young Gen Xers (those born in the late 1970s until 1982 and are in their late 30s to mid 40s) are the beginning of the second baby boom wave to hit the U.S. The boom occurred despite the roadblocks mentioned earlier, plus the fact that younger boomers were waiting longer to marry and start families. They are very different from their oldest siblings. The oldest Gen Xers (those born between 1966 and the mid-1970s and are in their late 40s and early 50s) are much more like their boomer elders than they are like their younger counterparts. In no other age group is this distinction so apparent.

The Generation X age group is more ethnically diverse than their older counterparts and are better educated. More than 60%

of all Xers have attended college. Gen Xers are very individualistic. They are the first generation where the term "latch-key kids" was used because they were the first to grow up with both parents having to work to support the family. Additionally, the Generation Xers were the first to see large numbers of their peers growing up in broken homes as the stigma of divorce began to be stripped away in society and the incidences of divorce rose. For those reasons, Gen Xers found themselves forced at an early age to become independent and resourceful.[11]

Because many, if not most, of the Generation X age group have had to be independent and resourceful since childhood, they are also more likely to resent having a superior stand over them or micromanage them at work. This can lead to Gen Xers having a certain dislike for authority and to expect – if not demand – their freedom in the workplace. They also value the responsibility that goes with that freedom. Additionally, because they lived through difficult times economically during the early part of many of their lives, Gen Xers are much less likely to be loyal to a single employer, and much more willing to move from job to job to get ahead. This willingness to move and begin new jobs or even new careers makes Gen Xers much more adaptable when it comes to change and even to alternative lifestyles.[12] They are more politically liberal than the older generations, more culturally diverse, more sensitive to cultural changes – and that has had an impact on the younger generations.

Generation X is the first to be reared during the United States' economic move from a manufacturing economy to a service economy. As such, Gen Xers are much more likely to resemble their younger age groups in enjoying and being adept with new technologies, even embracing them, than the baby-boomers and the mature generation. For Gen Xers, it's computers, cell phones, and satellite television, instead of over-the-air television, fax machines, and landline telephones.

Old Gen Xers still remember television without cable. The oldest Xers were nine years old at the time of the Ali/Frazier match on HBO, one of the two precipitating events that began the cable revolution. They remember the time before desktop computers, satellite dishes, "Pong," the Walkman, Simon electronic game, floppy disks, microprocessors, word processors, digital cameras, laser printers, and cell phones. By the time the young Gen Xers came along, all the aforementioned technologies were part of their world, along with the Cray supercomputer, MS-DOS, and the first IBM-PC. With their earliest recollections the young Gen Xers would remember the early Apple computers along with the early Macintosh computer, the Windows operating system, and the disposable camera. They would remember as youngsters the introduction of the first 3-D video game, the CD-ROM, digital cell phones, and even the earliest analog versions of high definition televisions. The Generation X age group has seen a tremendous amount of technological change in a short time.

Generation X watched television programs that entertained, but also, at times, challenged the programming of the earlier periods. For every *Happy Days* there was also an *All In The Family*. Television for the Gen Xers included programs that starred minorities – *Sanford and Son, The Jeffersons, Diff'rent Strokes,* and *Good Times*. They watched programs celebrating older adults – *Maude, Marcus Welby, M.D.,* and *Barney Miller.* Television included cohabitation (*Three's Company*), anti-hero heroes (*The A-Team*), and even programs with a spiritual focus (*Highway to Heaven*). And, of course, there was J.R. Ewing.

As mentioned earlier, the Generation X age group is unusual in that it is the true transitional generation of the groups. Older Gen Xers are, at times, more like the baby-boomers, while the younger Gen Xers are firmly in the mode of their younger counterparts, the Millennials and Gen Zs. The oldest Gen Xers are currently in their early 50s, while the youngest Gen Xers are in their mid-30s. In fact, as will be discussed later in the chapter, Xers are becoming more and more like their younger counterparts to the point that the Deloitte Corporation has coined a new term for the youngest three age groups – the MilleXZials. Given the changes in technology, science, and culture, it is easy to understand how this generation is the transitional generation, encompassing both the best – and the worst – of the older and younger generations.

The Millennial Generation

The Millennial generation is the largest of all the generations, having surpassed the baby-boomers in 2015. They are the generation born between 1983 and 1996, making the oldest millennials turning 36 years old as of 2019 and the youngest 23 years old. As the largest of the generations, they have and will have the most short-term influence on the 21ˢᵗ Century Television revolution.

The two youngest age groups – the Millennials and Generation Z are very different from the earlier age groups (although, as mentioned earlier, the Gen Xers – and especially the younger Gen Xers – are becoming more and more indistinguishable from the younger two age groups). According to a variety of research, Millennials are sheltered, very tech-savvy, pampered, and have been made to feel extremely special by their parents, to the point that the term "helicopter parent" has become a part of the language used to describe those parents. Millennials have a very different idea of what is acceptable dress, including believing that it is fine to wear flip flops to job interviews with Fortune 500 companies (even outside of Redmond, Washington, and Silicon Valley!). For the millennial, using technology is like breathing – it's a natural part of life. They are facile with all kinds of popular technology, easily embracing and having to have the latest "new thing" to hit the market. For millennials, the laptop and, more likely, the mobile phone – especially a multifunction smartphone – appear to be at the center of their daily universe because of their continuing love of personal music libraries and their constant need to stay socially connected with their friends, classmates or work colleagues, and almost anyone who comes in contact with them, online or offline through text messaging, instant messaging, and social networks. They enjoy taking pictures with a mobile phone or traditional digital cameras and posting the pictures for all to see on Facebook and other such social media sites. Millennials love videogames, whether playing them on traditional videogame consoles, on social media sites (such as Candy Crush and others on Facebook), or online, where they can compete with other gamers from around the world on an equal footing.[13]

The Millennial generation is the first generation to feel a real need to stay connected 24/7. Millennials by-and-large are texting fiends, and it's not unusual for them to send and receive thousands of text messages a month to and from others.[14] Additionally, virtually all millennials (94%) maintain a social networking site, again as a way to keep in touch with others in their ever-widening sphere of contacts. Millennials also are "LinkedIn," "Twittered," "Instagrammed," and "Snapchatted" using LinkedIn to help them network as they begin their careers, while the others are good for chatting about almost anything.[15]

For the millennial generation, television is anywhere and everywhere, with professionally-generated content (notice the change in the term for traditional television programs preferred by millennials) competing for their attention with user-generated content ranging from slickly produced programs to the most rudimentary self-aggrandizing fluff. Further, watching television can have absolutely nothing to do with any form of traditional television, but can mean watching streamed movies from Netflix or other such companies, watching Internet-only programming over set-top boxes such as Roku and Apple and Google TV. Television can also mean watching traditional television programs on their laptop computers, their smartphones, and, now even their tablet computers. Additionally, television can also mean making their own programming that they upload to YouTube, Vimeo, and other such sites, then watching their own or friends' programs as well.[16]

To millennials, technology, especially their media technology, is their theatre, and content of all kinds, professional-generated as well as user-generated, take center stage in their lives. They are the first generation to be truly active participants in their media and content experiences, and consider their own forms of content equal in interest and viewability as the professional content. They are the leaders of the 21st Century Television revolution occurring today, and will be the ones who will define the near-term future of what that television universe will be. Watch out traditional television – the Millennials are on the move!

Generation Z

Generation Z is the youngest of the television generations surveyed. They were born between 1997 and 2004, and range in age from 15-22 as of 2019. If the millennial generation was different, then Generation Z is truly special as they will be the beneficiaries as well as the drivers of the 21st Century Television universe. They have been called by others as the "iGen" generation – "i" standing not for iPod, iPad, iPhone, etc., but for the word "instant" in that they want everything NOW(!), and have no real understanding of the idea of delayed gratification.

The Generation Z world will be unrecognizable to the older members of the baby-boom generations, while the younger baby-boomers and Gen Xers will struggle to keep up with the two younger age groups. For Generation Z, the days of the slow and steady cycle of innovation from introduction of a new technology, to acceptance, to mainstream use of that technology have irrevocably changed. Gen Zers are capable of, and even embrace, the absorption of numerous technologies at the same time, and do so with a seemingly complete lack of effort. They are able to "multitask" with multiple technologies because so many of today's technologies have been invented just in their short lifetimes. They are growing up in a world where all the technologies of the 1990s are practically ancient to them, while the technologies of the 2000s are just their normal everyday playthings.[17]

Whereas the older generations were given dolls and cowboy hats as children, Gen Zers are being given electronic, computerized, learning consoles operated as simplified computers and toy cell phones. Since the turn of the millennium, they have seen, among other inventions, the iPod, iPhone, and iPad from Apple and the Android operating system from Google for a multitude of different devices, including mobile phones and tablet computers. They enjoy the updated game consoles with the ability to play DVDs, to connect to the Internet to play online videogames as well as console games, to rent movies and television programs from Netflix and other such companies, and be able to recognize movement. They interact with a multitude of social networking sites, including all the most recognizable ones. They listen to satellite radio and save their homework on a flash drive, the device that has replaced zip drives and CD-ROMs for the most part.

Generation Z is the first generation to be totally consumed with tweeting, skyping, and texting more than talking on the telephone. Telephone? Oh, sorry - they spend more time texting than talking on their mobile phones. Telephones, for the Generation Z age group, are those silly, antiquated things their grandmothers and grandfathers still have in their homes that are connected by a ridiculous wire, even if they claim them to be portable. Even their parents have gotten rid of their "landlines" (the modern term for telephones), unless the family lives in some out-of-the-way location where they still "can't get good signal" on the mobile phones. For anyone else, a landline is about as useful as a good doorstop.

For the Generation Z age group, television is – well what it isn't, is likely a better way of looking at the question. Television isn't just watching over-the-air programs using rabbit ears to get the signal, and, increasingly, it isn't even watching television delivered through cable or a satellite dish to the home at the time the program is actually telecast. Rather, for Gen Zers, they want their television and they want it available anywhere, which includes on the platform of their choice and available anytime 24/7. They aren't interested in appointment viewing, but ubiquitous video on-demand. They want their television to be interactive, including two-way communication (two-way? – no, every way!) and completely under their own control. For Gen Zers, television must be able to be shared, so that they have the ability for television to be a completely collaborative experience with their friends, relatives, and others no matter where in the world they are, through their social media accounts, mobile phone applications, and any other device that might be available now or in the future. In other words, they want it all and they want it **now!**

The Generations Today
(**Note:** Unless otherwise cited, all the information in this section comes from the three most recent editions of the Deloitte *Digital Media Trends Surveys*.)

"The movement of consumers away from traditional media viewing and toward streamed online content continues to accelerate, forcing companies across industries to plot new strategies." So began the 11th edition of the *Digital Democracy Survey* in 2016. In reality, this section could begin and end with that statement, especially for the youngest three generations – the ones that will most impact 21st Century Television. If the reader was to return to the first edition of the first book in this series,[18] (s)he would read where, for all the generations, traditional television would be the overwhelmingly preferred way of viewing television. Today, while traditional television viewing is still the most popular overall, television viewing online is closing the gap every year.[19] For instance, from 2016-2018, the overall percentage of U.S. households sub-scribing to at least one streaming service (not skinny bundle) rose from 49% to 69%, and, for the first time, overtook pay-TV services in the home.

However, the choice between pay-TV services and streaming services is not an either/or proposition. Indeed, 43% of house-holds have both, in the latest survey. That percentage, though, is mitigated somewhat by the fact that almost three-quarters of baby-boomers (74%) are still subscribing to pay-TV services while only half of boomers subscribe to at least one streaming service, the only group mentioned in this chapter that is more likely to have a pay-TV service than a streaming service. The other three groups choose to subscribe to a streaming video service more often than to traditional pay-TV services. Of the three younger groups, the Gen Xers are, once again, the transitional generation. They do prefer streaming video services over pay-TV, but the mar-gin is much narrower than the two youngest generations. Slightly more than three-quarters (77%) of Gen-Xers subscribe to stream-ing video services compared to two-thirds that choose traditional pay-TV services. The closeness of the spread suggests that a large number of Gen-Xers subscribe to both, but it is also reasonable to assume that – because of the transitional nature of this age group – the older Gen-Xers are the ones most likely to be subscribing to the pay-TV services, or to both video streaming and pay-TV ser-vices each month.

For the millennials and the Gen Zers, the spread is much wider: almost nine out of ten (88%) millennials subscribe to a video streaming service while only about half (51%) subscribe to pay-TV – the widest spread among the age groups, while eighty percent of Gen Zers subscribe to a streaming service versus 57% that subscribe to a pay-TV service. Given that the age group begins at 14 years of age, it would be interesting to break out the college-age Gen Zers who are most likely living away from home for all or a large part of the year from those that are in high school and still living at home full time. Because of today's expense for subscribing to a pay-TV bundle, the percent of those older Gen Zers subscribing to pay-TV likely is in single digits or even approaching zero, while the percent subscribing to streaming services likely is in the high-nineties and possibly 100% or close to it. Given the width of the spread between pay-TV and video streaming subscriptions, the number of new subscription services that are available or are coming online in the near future, and the number of major corporations that are beginning to develop their own "skinny bundles" of channels, the days of the traditional pay-TV subscription are almost certainly numbered as the baby-boomers and the older Gen-Xers age up and die out.

In addition, consumers in the four age groups are willing to spend and to manage their streaming video services. Across the age groups, on average, consumers subscribe to three different video streaming services per household. Interesting – but not surprising – when streaming, the oldest three age groups discussed here are more likely to stream using a paid, non-advertising streaming service like Netflix, Amazon, HBO, etc., than a free, advertising-driven service. Only the Gen Zers are slightly more likely to use a free video streaming service than a paid service, but the spread is miniscule (37%-39% paid service to free service). The other three age groups are basically identical (Baby-Boomers: 46%-29% paid to free; Gen X: 47%-25% paid to free; Millennials: 46%-28% paid to free). Also, skinny bundles do not appear to be of interest to any of the age groups with usage ranging from a low of 8% for boomers to a high of 11% for Gen-Xers and Millennials. This significant lack of interest in skinny bundles does not bode well for the corporations that offer these bundles and especially the traditional pay-TV services for whom the skinny bundle represents a last way to retain and/or attract subscribers. Additionally, the growth of streaming video services has led consumers to begin reassessing the value they are getting from their traditional pay-TV subscriptions. There is a widening gap between the value consumers expect to receive and what the pay-TV providers are actually delivering for the money consumers are paying each month.

While consumers enjoy their abilities to shape their content viewing with a la carte choices, they don't like the friction that comes with those opportunities. They don't enjoy the hassles, responsibilities, and even the vulnerabilities that come with having all the choices.

One of the top concerns for consumers across the age groups is when programs they enjoy disappear from the service they are accustomed to watching. Viewers of traditional broadcast television have had to deal with this problem for years when networks would cancel a favorite show. Streamers are also having this concern, but in two ways. First, for the first time ever, Netflix has begun cancelling shows and series that it had in its library.[20] Of greater concern for consumers, though, is the disappearance of programs as more and more studios and television networks pull their programming from the major streaming services to launch their own direct-to-consumer services. The streaming services that lose programming must either fill those holes with other, often original, programs, or have a library with fewer offerings and potentially less value. Consumers, then, have a decision to make – add the new service(s) that has the consumer's favorite program or do without. Both choices are less than optimal for the consumer.

A second concern for consumers is the frustrations they feel having to subscribe to a variety of different services to piece together a lineup of programs they want to watch. Forty-eight percent of consumers across all four age groups say it's harder to find content when it is spread across multiple services, while 43% say they give up on the search for their favorite content if they can't find it in a few minutes. Despite having numerous options to choose from, many consumers still feel that a good program is hard to find, according to the most recent survey. This concern will likely only increase as the many new subscription video services come online.

A third concern voiced by the respondents of the survey is the vulnerability to security breaches that they feel, along with the loss of privacy that's involved as they are required to provide more and more personal information every time they sign up for a service. Consumers across all the age groups, and especially the older Gen-Xers and baby-boomers, fear identity theft, financial loss, and the use of sensitive data without their authorization. As a result, the respondents want to control their data the way they are controlling their streaming experience. Forty-nine percent of the respondents on average across the various age groups believe they are responsible for protecting their data and 88% believe they should own their personal data. Interestingly, just seven percent believe that the government is responsible for protecting their data.

When it comes to monetizing the viewer, advertising – specifically, addressable, personalized advertising – is likely to become increasingly important. As has been pointed out earlier, there is a proliferation of SVOD services that are already available or will be available in the near future. Additionally, the consumer is unhappy when her/his favorite show gets pulled from one service's library and becomes a part of a different subscription service, thus making the consumer either purchase an additional service or do without that favorite program. As the costs of subscription services mount, consumers may actually find themselves paying as much, if not more, for their subscriptions each month than they did with their traditional pay-TV bundle. At that point – which will likely be reached sooner rather than later – the various streaming services themselves may begin to find ways to retain and attract subscribers by offering AVOD options, if they can provide the right amount of advertising per hour to satisfy their viewers.

So what is the right amount of advertising per hour that will satisfy the consumer? Interestingly, the age groups are actually very uniform in their preferences regarding the amount of advertising they feel is optimal per hour and what the cutoff number is for the maximum number of advertising minutes there are in an hour. Across the age groups, the right amount of advertising minutes per hour ranges from a low of seven minutes for the baby-boomers to a high of nine minutes for Gen Xers and Gen Zers. Further, the maximum number of minutes per hour that is too many minutes ranges from 15 minutes for the baby-boomers to a high of 17 minutes for, once again, the Gen Xers and Gen Zers. While it is often touted that it is the younger generations that don't want to watch commercials, it is the oldest generation – the baby-boomers – that are the least enamored with watching commercials. Perhaps that is because they have seen at least 100s of thousands of commercials throughout their lives!

Consumers are willing to trade personal information in exchange for personalized information, or information that delivers an equal or superior value to them for that information. The success of this personalized, targeted advertising depends heavily on the ability of the advertisers to gain insights into consumers' demographics and viewing behaviors, augmenting that information with social media data, and – when the consumer is watching on the go – location data. However, the willingness of the consumer to share this data is critical. The good news is that the U.S. consumer still wants value-added services from companies such as premium content, personalized experience, and ads, discounts, and rewards. Fortunately, addressable advertising (especially as discussed in the coming chapters of this book) can, does, and will provide those very things that the consumer wants for his/her data. In the 12th edition of the Deloitte survey, researchers found that consumers are willing to share various aspects of their personal data, including

- Date of birth
- Video viewing history

- Purchase data
- Contact data
- Browsing history
- Online search history
- Social media activity, and
- Home address

That list provides a rather substantial amount of in-depth information about the consumer and, when augmented with profile data, ratings (if they still exist), and marketing data that can be purchased from a variety of companies each with a diversity of information, will make it possible to give the consumer the personalized advertising and experiences (s)he desires. If the advertisers and the content providers can give the consumer that personalized advertising and experiences (s)he desires, along with providing the protection of the consumer's personal information, and keeping the number of advertising minutes to a reasonable level, then the consumer will lavishly embrace the possibilities that this new type of advertising will afford her/him.

Summary[21]

21ˢᵗ Century Television has already seen, and is continuing to see, dynamic and monumental change in the television viewer. While the 20ᵗʰ century saw the development of the television viewer into a sophisticated observer of content and advertising, 21ˢᵗ Century Television is facing a content and advertising savvy and demanding viewer.

The first quarter of the century will see the "torched passed to a new generation"[22] with the Baby-Boom generation giving way to the Gen Xers who will lead the continuing revolution that is 21ˢᵗ Century Television. It will also be a time when the Millennials and the Gen Zers will be preparing themselves to assume the mantle of television leadership, both from the professional side and the consumer/viewer side. They will be the driving force behind the changes delivered by Gen Xers that will change the way television is delivered and monetized.

The second quarter of the century will see the millennial generation well entrenched as the powerhouse generation for 21st Century Television. By this time, television will be unrecognizable from the television the baby-boom generation grew up watching. Television will be well on its way to moving from viewing to experiencing, and the monetization of television described in these pages will be normal and ordinary. Viewing will be anywhere, anytime, on any platform, and television will be entering an era unlike any seen before.

The second half of the century will have Generation Z in charge. They will revolutionize 21st Century Television once again, fully meeting all the expectations of 21st Century Television. By the second half of the century television as an experience will be a given and will be able to be enjoyed as an experience anywhere and anytime. Platforms likely will not matter as delivery mechanisms will most probably be as much internal as external – Gen Zers will make sure of that as they prepare the next generation – their daughters, sons, and grandchildren – for whatever the next revolution in television will be as they move from 21st Century Television to 22nd Century Television (if "television" is even a term any longer). For the television consumer (rather than just viewer) the coming years of 21st Century Television will be an amazing transformation, one not to be missed.

PART II:

MONETIZING 21st CENTURY TELEVISION

Section 1:
21st Century TV Advertising

CHAPTER 4
ADVERTISING - OVERVIEW

2$1^{st}$ Century Television will be significantly different from today's television. Today's (2019) television is – for the most part – little changed from television in the 20th century. In fact, in terms of television advertising, almost nothing has changed since television was first introduced in the 1950s (when television became a household product) except the length of the commercials and the sophistication of the creative aspects of the production (with notable exceptions).[1]

However, as television progresses through the 21st century, from the legacy media delivery systems to IPTV, VOD (and all its subcategories), cloud storage and delivery, the numerous platforms on which television will be available, and even the alternative delivery systems that are on the near horizon (such as ATSC 3.0), television advertising itself must change. No longer can television professionals hope to use the techniques of today's television to reach the vast and fragmented audiences of 21st Century Television. The antiquated ways and methods of legacy media advertising must change or those media will be replaced, to the detriment of the viewers.

Fortunately, though, the ideas of how to advertise to the 21st Century Television viewer are slowly changing and evolving as new companies develop and make available new tools for monetizing 21st Century Television. While the broadcast media – especially the local stations – are most likely to cling to the traditional, though archaic, forms of advertising delivery, the cable and satellite industries, and even the broadcast networks are looking for answers to monetizing their products. Whether it be ATSC 3.0, expansive retransmission consent agreements (which, as discussed in previous books in this series,[2] will ultimately disappear), or 21st Century Television advertising, the legacy media continue to look for ways to improve their bottom lines.

As a reminder, since the late 1940s, American viewers have been continually conditioned to the idea that television will have commercial breaks in the programming. Initially, commercials generally ran for one minute at a time. As television advertising progressed and viewer sophistication with television advertising grew, the one minute spot gave way to the classic 30-second spot, which dominated the television advertising landscape for most of the history of the industry. Even at this time, the 30-second spot lives on, although – once again because of increasing viewer sophistication with television advertising – more 20-second and even 15-second spots are airing on programs. It is now possible to tell the advertising story in as little as one-fourth the time that it took to tell the story when television was young, thanks to the ever-increasing sophistication of both the advertising creative professionals and the viewing public.

Why, then, are television executives, especially broadcasters, so certain that the public will not watch commercials on television programs viewed over the Internet? After all, the broadcasters and advertising professionals have worked so hard and so diligently for more than half a century to make sure the viewing public is conditioned to watching as many as 64 commercials an hour. (Assuming all commercials during the hour are the 15-second variety – 64 commercials would be the maximum number that could run.) Why buy into a notion that today's advertising methods will be tomorrow's advertising methods and that 21ˢᵗ Century Television viewers will only watch television through legacy means – a notion that is completely illogical and counter to everything that the broadcasters and advertising professionals have worked so hard to perfect since the late 1940s? It makes no sense, and the longer broadcasters hold to the notion that viewers will not watch commercials (or minimal numbers of commercials at the most) in programming delivered over the Internet, the more they reinforce that notion until it potentially becomes a self-fulfilling and, ultimately, a self-defeating prophecy.

Instead, broadcasters and all legacy media content providers should fully and completely embrace the Internet for the delivery system it is, and make good use of the unique capabilities for advertising that the Internet provides. Viewers conditioned to watching television in the traditional ways will watch television the same way over the 21ˢᵗ Century Television delivery system of the Internet. Younger viewers must also be conditioned to watch commercialized television over the Internet. Otherwise, they will be the generation to begin the self-fulfilling prophecy of advertising as unacceptable on television. Fortunately, AVOD (advertising video-on-demand) is beginning to take hold, with Amazon being the latest entrant into the AVOD business with its launch of IMDb Freedive, an ad-supported service. IMDb Freedive joins, among others, Tubi, Crackle, Pluto TV, Wurl, Inc., Xumo, and The Roku Channel.[3]

Whether the content providers like it or not, television will move to the new delivery system of IPTV (or some other form of Internet-delivered television) and commercials will continue to air in programs as they have – although the way the advertising is structured, developed, designed and delivered will be vastly different.

This is what is meant by the title of this book, "Monetizing 21st Century Television."

So how will advertising work on 21st Century Television? That is the topic for this chapter and the next three chapters. In reality, it will be more efficient and more effective than on television today. Traditional television advertising has been a shotgun approach – you advertised to the largest number of people and hoped that a reasonable number of viewers responded to the advertisement. If they did, the ad was effective. It not, well, you tried something else. While there is always talk about the target audience, the target market, and advertising to specific demographic and psychographic groups, in reality, advertising on television still remains a shotgun approach because it is simply not possible to target more narrowly on traditional over-the-air, broadcast television.

For cable and direct-to-home satellite services, it is easier to target specific groups because so many of the channels are niche-audience focused. However, even on those most niche-oriented channels, targeted advertising is not nearly as specific as it could be. Most cable- and satellite-delivered channels see their viewers as aggregate audiences even if they are more tightly defined, at least on their traditional cable delivery systems. While, say, the History Channel may target history buffs, or ESPN might target football fans in the fall, basketball fans in the winter and early spring and baseball fans in the summer, those groups are still lumped into single aggregate groups and only rudimentarily broken out any further. Nevertheless, the examples of the niche audiences on cable and satellite TV today begin to demonstrate the possibilities for 21st Century Television. Because the programming is already niche, the advertising can also be niche – but not to the degree 21st Century Television will offer.

Then there is the question of interactivity. Traditional television is simply one-way television. Television advertisements are delivered to the audience with almost no way for the audience to respond directly to the message. The communication is one way, not two way. Cable has the ability to deliver interactive, two-way advertising, but, for the most part, it has not done so. Traditionally, cable television channels have taken the same approach that the broadcast television channels have taken – delivering their advertisements in a one-way manner. Whether it is the broadcast channels or the cable specific channels, one-way advertising – as well as one-way programming to their audiences – is the time-honored tradition. The two-way programming that has occurred on television and more often on cable is in the form of home shopping networks, infomercials, and hard-sell advertisements that request immediate 1-800 telephone call orders, or in the form of certain reality programs where the audience is urged to call, text, or log on to the channel's website to vote for a favorite participant. While these are all – in a sense – interactive, in reality, they are a poor beginning for what is possible with true, interactive television advertising.

As far back as 2011 or 2012, some of the more forward-thinking cable system owners, especially those such as Comcast and Time Warner, were beginning experiments with interactive advertising.[4] Cablevision added an opt-in interactive option to its cable services called Optimum Select, where viewers could request coupons, more information about a product or service, or other longer-form content.[5] Such experiments, if ever successful and fully implemented on a full cable schedule, are the forerunner of the truly interactive advertising that will be available on IPTV.

Twenty-first Century Television, delivered as IPTV, provides a whole new level of advertising possibilities that are either not available, or that are more available, than what advertising is currently on legacy television. In fact, some have suggested that we could see not only completely interactive television commercials, but also true microadvertisements, or advertisements literally targeted to an individual. The first is certainly possible; the second is, most likely, not economically feasible, for the reason that it is not pos-sible to make millions of different commercials to deliver to mil-lions of different viewers on a true, one-to-one basis. However, the technology for true microadvertising does exist and is available today and early experimentation is occurring.[6]

There is, though, an alternate form of advertising, using cer-tain aspects of microadvertising, that will be economically feasible as well as extremely desirable, and will be discussed later in the next three chapters. This alternate form of microadvertising will require a three-step process, each step explained in those subse-quent three chapters.

Total interactivity will be a mainstay of IPTV. With a click of a remote control or a mouse – or even a voice command, through the use of Amazon's Alexa or Google's Google Home (the two best-known of such systems), or a wave of the hand – viewers will be able to change views of the product, change colors, open win-dows to provide additional product information, find locations to purchase the product, etc. Additionally, depending on the specific product, interactivity will include the ability to make an immedi-ate purchase, upgrade an existing product, extend warranties, etc.

Twenty-first Century Television, because it is delivered over IPTV (whether through wired connection, Wi-Fi, or mobile), will operate much like what is already being delivered on trial bases over computer websites today. With the connected televisions that set manufacturers provide to users, with separate windows on the screen for widgets, chat rooms, websites, etc., the ability to have total interactivity is available to advertisers. They just have to deliver the commercials that have interactive elements in them, so that the viewer can make decisions immediately. Connected tele-vision sets are currently available from virtually all the television set manufacturers, complete with software from Rovi, Google/ Android, Roku, and other software makers that make the multi-screen, interactive television sets possible.[7]

Compared to the sets earlier this decade, the latest connected television sets are even more impressive than their predecessors. Even so, they are still extremely basic to the television sets of the next few years and into the future. Nevertheless, the hardware and software are moving closer to what will be necessary to produce the interactive television set of the future. Such interactivity will provide not only a much richer purchasing environment over the television, resulting in more impulse sales of products and services, but will also provide the advertising agencies and their clients with continuous feedback on the success or failure of their advertising campaign creatives.

What is most important to advertising agencies and their clients, however, will be the ability to target much more narrowly their chosen audiences and markets. The ability to narrowly target these specific audiences will allow advertising professionals and their clients to have maximum impact with their commercials and maximum success with their campaigns. Targeting the audiences will be much more specific than even what occurs in cable and satellite niche programming today. The idea of narrowly targeting the audience is often referred to as "addressable advertising." Addressable advertising is defined as "the ability to show different ads to different households while they are watching the same program. With the help of addressable advertising, advertisers can move beyond large-scale traditional TV ad buys, to focus on relevance and impact."[8] Gartner Research defines addressable advertising as "technologies [that] enable advertisers to selectively segment TV audiences and serve different ads or ad pods (groups of ads) within a common program or navigation screen. Segmentation can occur at geographic, demographic, behavioral and (in some cases) self-selected individual household levels, through cable, satellite and Internet Protocol television (IPTV) delivery systems and set-top boxes (STBs)."[9] Addressable advertising is the way in which 21ˢᵗ Century Television advertising will be delivered and by which

21ˢᵗ Century Television viewers will receive their advertising. It will be targeted to them, delivered to them with exactly the advertising they want, in the manner in which they want, at the time they want, through the platform they want – anytime, anywhere, anyway.

To accomplish the goals and abilities of addressable advertising, the process will occur in three steps. The first step in the process is *Aggregated Targeted Microadvertising*, or ATMA. Aggregated Targeted Microadvertising provides the mechanism by which audiences are segmented so that each segment can be delivered advertising that is right for them – the proper products and services, the proper creatives, the proper messages, designed just for that specific audience segment and delivered in the way most likely to produce a winning response.

The second step in the process is *Programmatic Advertising*, or, more properly, *Programmatic Media Buying*. Programmatic Media Buying takes the audience segmentation results produced by the use of ATMA and first matches the segments with the current programs those segments regularly enjoy. Next, Programmatic Media Buying matches the audience segments with programs that are similar to those current programs both in the past inventory of programs the content provider has available as well as new programs the provider is planning to introduce in the near future. The final step is the purchase of advertising time in those specific programs for each audience segment.

The third step in the process is *Dynamic Ad Insertion*. According to the Interactive Advertising Bureau, dynamic ad insertion is "The process by which an ad is inserted into a page in response to a user's request. . . .At its simplest, dynamic ad placement allows for multiple ads to be rotated through one or more spaces. In more sophisticated examples, the ad placement could be affected by demographic data or usage history for the current user."[10] The Interactive Television Institute's *itv dictionary*, says about dynamic ad insertion, "With dynamic ad insertion, network operators can provide targeted ads that can be swapped in and out of that television program as it's delivered to the enduser."[11] Of the three steps in the process, dynamic ad insertion is the oldest, as it was used as far back as the earliest days of audio/radio streaming to denote the idea of placing advertising into various streams for individual listeners, without any real structural thought to the other two steps in the process. In terms of use, development, and understanding, DAI is currently the best known and understood, while ATMA is the least understood and in its most nascent development, with programmatic being somewhere in between.

Each of three steps in a successful addressable advertising process is discussed in much deeper detail in the coming chapters, beginning with Aggregated Targeted Microadvertising, then Programmatic Advertising/Media Buying, and finally, Dynamic Ad Insertion. The combination of ATMA, Programmatic, and DAI will lead 21st Century Television into a new "golden age" of television advertising, producing a "win-win-win" scenario for the advertiser, the viewer, and the content provider. The result will be a much richer television industry, a much richer advertiser, and a much happier and more satisfied television viewer.

However, the monetization of 21st Century Television does not stop at just the development of a new and innovative process for advertising. In addition, 21st Century Television viewers will enjoy the results of the development of *Ubiquitous Product Placement,* or UPP. Ubiquitous Product Placement takes the product placement occurring today on television and expands the possibilities for viewer interactivity to every frame of every program the viewer chooses to watch. UPP adds a second opportunity for advertisers to reach consumers by making the opportunity for an "impulse buy" possible at a moment's notice while the viewer is enjoying her/his program.

Television relies on having viewers watching their programs – 21st Century Television will be different in that respect. However, as has been discussed in the previous two sections, finding ways of reaching, exciting, and connecting with viewers will be even more critical in a VOD, any-platform television world. One of the most important aspects of 21st Century Television is the need for continuous promotion by the content providers. With the demise of schedules brought about by the growth of video-on-demand, making sure that audiences can find the programs they want to see will be a critical priority for the content providers. VOD will erase the "appointment viewing" so popular on all of the legacy media choices today due to the fact that viewers must accept the media's scheduling of programs. With 21st Century Television, there will be no new seasons, no 2nd seasons, no summer reruns. There will be no starting times and no ending times – except for sporting and other live events. Even so, viewers will be able to watch the various live events on their own time, not on the programmer's schedule. Further, because advertising will be individualized, promotion will be critical for driving audiences to the programs where advertisers can successfully reach them. Promotion, then, will be critical for the successful monetization of 21st Century Television.

Finally, 21st Century Television will be global television, with U.S. programming reaching into every corner of the world and the world's television reaching into every corner of the U.S. As such, the monetization of global television will occur, providing new opportunities, new products and services, and new viewing audiences for the 21st Century Television content provider. The globalization of 21st Century Television will be the final step in the 21st Century Television evolution and will make possible an era of prosperity for the television industries the likes of which has not been seen.

CHAPTER 5

AGGREGATED TARGETED MICROADVERTISING

Aggregated Targeted Microadvertising (ATMA) is the first step in producing advertising for 21st Century Television. ATMA can be defined as "the process of determining large subsets of an audience for any particular television program, based on numerous specific characteristics that are inherent in every member of the subset, then developing an ad or a series of ads tailored to meet the needs and desires of each particular subset." (NOTE: This is the author's term and definition.) Given that, during each television program more than one subset is likely to be identified, an ad for any particular subset would be specific to that subset, with a different ad being delivered to each subset. The result is that the total number of ads running in each commercial spot time period will equal the total number of viewer subsets watching the program.

The author chose the phrase aggregated targeted microadvertising because at this time there is no one term generally accepted by the television or advertising industries for this first step in the addressable advertising process.[1] Today, television professionals use (or, rather, misuse) the term dynamic ad insertion, or DAI, for both what DAI actually covers as well as for ATMA. This book, however, more correctly separates the two terms and adds in Programmatic Advertising/Media Buying, using each of the three terms to identify what is the correct function for each step in the whole addressable advertising process.

ATMA is not specifically microadvertising. Microadvertising is advertising that directs a different ad to each person, so for one million people an advertiser would need one million different advertisements. In ATMA, the key word is "aggregated." While microadvertising is possible, but is neither feasible nor economically viable, aggregated targeted microadvertising will be the way of the future. Early generations of ATMA were being developed and introduced to television as of mid-2012, and today, there are strong, but still fledgling Programmatic and DAI industries for 21st Century Television. The major impediment to full adoption is the concern by the legacy media to make the changes in delivery systems necessary for addressable advertising to be successful on a macro scale.

ATMA makes possible the delivery of advertising to specific individuals much like microadvertising, but, unlike microadvertising, it uses only a set number of advertisements and directs one of those several advertisements to each specific individual in the targeted audience. There continues to be small-scale (comparative to traditional advertising methods) usage of alternative advertising models on websites like Hulu and the legacy network websites such as NBC, where they provide the viewer with two advertisements and let them choose which of the two ads they wish to watch. As far back as February 4, 2010, an article in the *Wall Street Journal* reported that research demonstrated that when viewers were allowed to choose from among three advertisements which one they wanted to watch, the consumers were more likely to watch the advertisement, recall the advertisement they watched, and were more engaged with the advertisement. The research study, conducted by a subset including the advertising agency Publicis Subsete, Microsoft, Yahoo, CBS, and Hulu, also showed that choice was preferred by consumers over approximately 30 other forms of advertising formats currently in use.[2] While the results showed choice to be preferred even over interactive online videos and clickable videos, it must be remembered that only the choice format allowed the consumer to have his/her preferred advertisement.

How ATMA Works

As mentioned earlier, the key word in ATMA is "aggregated." For any program there are multiple subsets of viewers in the audience who are different from each other in certain aspects and alike in certain aspects, but who intersect at the program. To begin the ATMA process, the various subsets of the audience watching the program who are interested in a particular product are segmented from each other by their demographic and psychographic details, along with all the additional information gleaned from their purchasing patterns, their geographic locations, their social media use and habits – any information that allows the advertiser to understand as fully as possible what is needed to reach each subset successfully. Each of these subsets is studied carefully to determine those aspects of the product and commercial creatives that are successful at connecting with the subset members and are the most likely to be persuasive for that subset.

Once the agency has developed those creative aspects that will make each subset highly likely to identify with, and be persuaded by, a particular type of commercial, the advertising agency then produces possibly a variety of different commercials, the number of commercials that must be made being dependent on the makeup of the subset. Each commercial need not be completely new and different from all the others. Quite possibly, the differences will be as slight as the person selected as the spokesperson for the product. In one ad, a young woman might be featured; in another, a young man. A third ad might have an older woman, and a fourth, an older man. A fifth might have a couple; the sixth, a family. Other slight changes could be in the race, the ethnicity, or the lifestyle preference of the spokesperson(s). Depending on the specific subset, the setting might also change. One commercial might be set at a sporting event; another might be on a picnic. A third commercial might be set on a plane, while a fourth could be set at a concert. Additionally, because products almost always have at least one secondary target audience, the selection of the individual might appeal to the primary target audience subset, while the setting could appeal to a secondary audience subset, or, possibly, the other way around, depending on which aspect would be the most powerful for reaching the primary target audience subset. The research (discussed later in the chapter) would be designed to let the advertiser know what choice(s) would be most likely to be successful to reach each portion of each subset interested in the particular product.

Due to the necessity to produce a number of different commercials, even though the costs of production would likely be higher than today's style of advertisements, the ability to reach the target audience with advertisements that will have a very high degree of acceptance and response by the members of each particular audience subset will more than pay for the extra costs associated with the creation of multiple advertisements. Even the youngest of viewers choose to watch advertisements that they want to watch – ones that have their favorite products; their favorite stars; their sports, music, etc., idols; and other such potential product spokespersons. In addition, in each case, there will likely be one or more scenarios that appeal to each of the particular groups of that target audience subset. Combining these two elements with others learned through the research makes each commercial a powerful delivery mechanism for a product.

After segmenting the audience and preparing the advertisements designed to reach each subset efficiently, the programmatic media buying process begins. Each subset will be linked with particular programs that the subset members are attracted to through the use of the programmatic media buying software. Once the links have been developed, the programmatic media buying software will place orders for commercial time in those programs. The commercials will then be placed on the various streams through the use of dynamic ad insertion, with each commercial delivered directly to the subsets of viewers for whom that ad will strike the most responsive chord and will lead to the most likely possibility of a positive outcome. At any given point, all of the different advertisements might be airing at the same time on the same program, but each viewer subset gets only the ad that is right for those members. Further, because all programming will be delivered completely through video-on-demand, those ads will run for the viewer whenever the viewer is watching the program. Additionally, should the viewer watch the program more than once, the advertiser has the opportunity to have a second, a third, or additional impressions from that one ad, or could rotate a series of ads if the repeatability of the program viewing is high.

At the same time, additional subsets of viewers watching the particular program but who are not interested in the product mentioned above will be receiving advertisements for their preferred products, with the creatives that will be most successful. As such, for each commercial time slot (30 seconds, 20 seconds, 15 seconds, etc.) every subset of viewers of a particular program will be receiving a commercial for the product of interest to them in the creative way that most appeals to them. Therefore, for every commercial timeslot there will likely be several different commercials for several different products streamed simultaneously to the various subsets. In that way, content providers will be able to sell the same timeslot multiple times to different advertisers, each reaching the audience subset they desire. The resulting revenue from the numerous simultaneously-sold commercials for each timeslot will more than equal the amount of today's one-ad-for-one-timeslot commercial delivery of television advertising.

The result of the addressable advertising process will be audiences for a particular program receiving advertisements throughout the program that are of the most interest to them in the way that will be the most persuasive. Instead of today's "shotgun" approach where a monumentally successful advertisement persuades 20% of the audience (or less), the addressable advertising process will produce advertisements that are successful at a rate of 80% or greater virtually every time. Further, with the ability to reach an audience again and again if they choose to repeat the episode through VOD, advertisers will have multiple opportunities for successful impressions.

Benefits of ATMA

ATMA provides the content provider with a number of benefits. The first benefit is that, unlike current television where advertisements can be skipped, especially in recorded-and-viewed-later programs, ATMA advertisements will be embedded so that they cannot be skipped, nor can they be changed, fast-forwarded-through, or any of the other capabilities that are currently possible. It is possible that someone will leave the room, or begin focusing on something else while waiting for the program to resume. After all, all the various age groups discussed in Chapter 3 know how to, and regularly do, multitask – the younger age groups multitasking constantly while watching TV. However, because the advertisements are ones most likely to strike a chord with the viewer, (s)he is much less likely to have a desire to skip the advertisement. Further, because they are viewing ads for products they like and use, with the creatives they most enjoy, it is anticipated that much – if not most or even virtually all – of the multitasking will be for the products they are viewing in the advertisements. Likeability brings on interest, interest brings on desire, and desire brings on response. Having the product the subset members like, along with the creatives they enjoy, virtually guarantees a high percentage of success.

A second benefit of ATMA is that it can be used throughout all of the viewer's program choices. Because the viewer is targeted by the advertisement, it is the viewer that can be sold, not the program. In fact, advertisers would have both the opportunity as well as the ability to place advertisements on all programs watched by the viewer. Additionally, because every time the viewing subset selects a program to watch, and that information is correlated with all other programs of a similar nature and promoted to the subset,[3] it will be possible to know what programs the viewing subset is watching at any time, plus what programs they are likely to watch. Those programs could then be sold as a package to the advertiser, knowing ahead of time that, because of the profile built on the viewing subset, the members are extremely likely to watch those programs and be able to be successfully advertised to with the products the subset is most likely to be interested in and likely to purchase.

A third benefit of ATMA is that it can be combined with complete interactivity within the advertisement. At any point during the advertisement the viewer will be able to click anywhere on the ad to go to the product's home page, or – better yet, because the viewer is already likely to be familiar with the product – to go directly to a product purchase homepage either to buy immediately or to find a location where the product can be purchased. For new products, it would still be possible to go directly to the product homepage so the consumer could be provided with the information that (s)he would need so as to make a decision on whether or not to purchase the product. The ability to have complete, direct interactivity of products specifically chosen for the individual viewers would provide the advertiser with the opportunity for significantly more impulse buys than having to find the product online through a separate search, or having to find and then travel to a location where the product was being sold.

A fourth benefit to Aggregated Targeted Microadvertising is the ability to advertise to consumers around the world.[4] With the ability through the Web to make programming available anywhere a viewer is located, television programming over IPTV would no longer be confined to over-the-air signals, signals sent to a local cable system headend, or signals able to be received from a low-earth-orbit satellite that would be sent directly to the home through DTH satellite services. Instead, using the Web, television programs could develop worldwide followings, and be able to have worldwide advertising available for all advertisers who choose the worldwide capability.

Global brands would become stronger and more recognizable, and brands desiring to go global would have the opportunity to have the influence of television lead the effort in creating that global reach for those specific brands. Further, ATMA could be used to reach truly global, but highly specific audiences, because profiles would be able to be built for viewers regardless of the country or part of the world. Wherever a viewing subset lives, if there is an advertisement the subset enjoys for a product of their interest, the subset – no matter where those individuals that make up the subset might live, what their ages may be, or what their likes and dislikes may be – will watch the advertisement. While it may cost more in production expenses to make advertisements in other languages or with spokespeople from other countries, the choice to advertise globally would be the decision of the advertiser as would be the additional revenues that would come from advertising to those highly targeted global audiences.

A fifth benefit of ATMA is the ability to deliver different ads at the same time to different viewers of the same program. It is almost inevitable that programs will cross profile types, especially those programs currently telecast by the networks that – while somewhat targeted to a particular audience – garner very broad and largely undifferentiated audiences (hence the term "broadcast"). Because ATMA is defined by the individual viewer, aggregated by similar likes and desires, it is almost certain that there would be multiple viewers who would like the same program, but who would not all like the same product or advertisement.

Using ATMA, the content provider delivering the program could stream a variety of advertisements to the various consumers who were all viewers of the program, but not all alike in what they wanted advertised to them. While, say, a young male is watching a particular MTV program (as an example), a young female might also be watching that same MTV program. The preferences of the two viewers might intersect at the program, but not when it comes to advertising interests. Therefore, the young male viewer would get an advertisement directed at his interests while the young female viewer would receive an entirely different commercial for an entirely different product during that same commercial break.

On a wider scale, the same thing could occur with two different viewers in different countries – for example, one in the U.S. and the other in France. The two might intersect at the program and at their likes so that the same commercial – with or without translation, depending on the profile – might be aired. At the same time, the two might intersect at the program, but have different profiles for advertisements. At that point, two different advertisements would be aired, possibly for the same brand, or for the same type of product but not the same specific brand, or there could be two different advertisements aired for two entirely different products. ATMA makes it not only possible, but even desirable to sell each advertising availability (commercial spot) multiple times to multiple advertisers. This ability makes possible the opportunity for content providers to increase their revenues of today by multiples of what they receive for an ad insertion at this time due to the additional value each commercial will have because of the high rate of success for each commercial. This "premium" high percentage rate of success would allow the content provider the opportunity to sell each commercial timeslot for an aggregate amount that would likely far exceed the revenue that a content provider receives today for a single advertisement in a single commercial timeslot going out to an undifferentiated audience.

A sixth benefit of Aggregated Targeted Microadvertising is that it can be placed in every program in the content provider's library of programs. Video-on-demand allows a viewer to watch whatever program (s)he pleases, whenever (s)he pleases, and now, wherever (s)he pleases, on whatever platform (s)he pleases or is the easiest or most convenient. As such, the viewer will have the opportunity to enjoy any and every program the content provider has in its inventory. Because of the ability to enjoy every program in the content provider's inventory, ATMA makes possible the opportunity to reach the viewer with advertisements that (s)he wants, no matter what the program is, be it the latest new program release or a program from half a century before. Because the unit of sale for the content provider is the viewer and not the program, every program that the viewer watches will have the opportunity to have advertising in it that will be successful. Therefore, all programs – even old or unsuccessful/no-longer-successful programs – will again have value.

The advantage of Aggregated Targeted Microadvertising is the ability to target directly the viewer with an advertisement that has a high probability of success, the ability to reach different viewers with different advertisements that will have high probabilities of success with each viewer, and the ability to reach different viewers around the world with advertising designed specifically for each of them, all in an aggregated, and therefore, cost-effective way. Given the capabilities of 21st Century Television, Aggregated Targeted Microadvertising will be a highly successful, highly cost-effective way of delivering advertising to viewers of all types anywhere in the world in a fully interactive way.

Reaching the 21st Century Television Viewer

One consideration is how the advertiser will know what type of advertisement each viewer will like. One of the advantages of 21st Century Television is that tracking the audience in detail is now possible. With 21st Century Television, a viewer will go to the website of the program (s)he wishes to watch. Every time the viewer watches a particular program on the content provider's website, that information will be stored automatically by the content provider for future use. The first time a viewer goes to that particular website, a profile will be set up for the viewer. The content provider may ask the viewer to fill out a survey, or more likely, will ask the viewer to fill out a profile page for the content provider so that it can assist the viewer in providing the right type of programs and information (s)he would find interesting or entertaining. Every time the viewer returns to the content provider to view a program, (s)he would log in to the site through the site's authentication program – the log-in is free – and when the viewer selects a program, that additional program viewing choice would provide more and more information about the likes and dislikes of the viewer. As the profile of the viewer becomes more and more detailed, the content provider would be able to deliver suggestions to the viewer of new or different programs that fit with the viewer's profile.

Because the content providers have a number of programs that are similar in style, actors/actresses, genres, etc., even with the initial selection of a program from the inventory of programs, the content provider will be able to begin the recommendation process. Just as Amazon.com provides recommendations for virtually any product that a shopper might have shown an interest in, or when (s)he makes a selection, so too will the content provider be able to do as well. This form of "push" programming will drive viewers to more and more programs they will find of interest, with more and more commercials that are selected for them based on their ever-growing profiles.

Because the content provider would have a library of programs – some, in the case of the networks, that date back sixty years or more – the content provider could introduce the viewer to more than just new or current programs that would fit the viewer's profile. The content provider also would be able to "introduce" the viewer to a whole generation or more of programs that the viewer may have only heard older people speak of if (s)he was younger, or to "re-introduce" the viewer to programs (s)he may not have seen since the viewer was a child or a youth. Older shows could be revived, new shows would be able to be narrowly targeted to those viewers most interested in watching the programs, and every time any program was viewed, advertisements would be embedded for products that would correspond with the viewer's profile.

A second way the profile of the viewer will be built is through social networking.[5] As an example, already, content providers are building Facebook fan pages. These Facebook fan pages encourage their fans to post about the programs they are watching as well as encouraging the fans to get their Facebook friends to "like" the content provider's Facebook page. Further, because Facebook friends can observe the profiles of their friends, the content provider is able to learn much about the fan by checking the fan's profile page. Such additional information allows the network to further refine the profile of the fan and to be able to target better both programs and advertisements that the fan would be interested in watching. For fans who were already viewers of the content provider, the profile pages would be additional information used to better develop a more complete and thorough profile of the viewer.

Additionally, these fan pages of the content provider would offer opportunities for fans to rate programs, see promos for new, existing, and older programs tailored to their profiles, and participate in contests based on the programs – the opportunities to drive fans of the content provider's Facebook fan page are numerous. Facebook fan pages could also be used to introduce fans to the various actors and actresses that star in the programs the content provider would have in their inventories for the fans to watch. These introductions would allow the fans to become even more interested in the actors/actresses and thus the television programs they star in, which would then further encourage the fans to "check out" the programs. By building interest in the actors and actresses, the content provider would be enhancing its own closeness with the viewer, giving the viewer even more reason to return again and again to watch the content provider's programs that are being offered to her/him with his/her profile in mind.

By providing profile information on the various actresses and actors to their Facebook fans, the content provider also will provide a means for their fans to access those actors/actresses Facebook fan pages. There the fan could learn even more about the actor/actress, gaining background information about the star, his/her likes and dislikes, what other programs (s)he has been a part of (thus building interest in other programs the content provider might have in its library), what the star thinks about the current program (s)he is in – anything that might heighten the interest of the fan in the star and thus in the program.

While all this is going on with the Facebook fan pages, the content provider is collecting more and more data about each individual fan, deepening and enriching the fan's profile, giving the content provider more and more information to build the profile of the fan. Ultimately, the content provider will have such an in-depth understanding of the fan that it can drive programming and advertising to her/him that the content provider can be virtually certain the fan will watch and consume. The fan then becomes a viewer and a consumer of the advertising on those programs.

While all this is occurring with Facebook, the same is being done on Twitter, Instagram, Snapchat, and all the other social networking websites available to the consumer. Every Twitter tweet provides information about the viewer. Every status change on Facebook or any of the other social network sites provides additional information about the person who posted the information. Social media is one of the most powerful forces available to the 21st Century Television content provider in understanding in-depth the viewer as well as the potential viewer, and in developing a rich enough profile of each viewer that the content provider can deliver the exact programming that will be effective in capturing and holding the attention of the viewer.

So where does this leave advertising? By developing such a rich and in-depth profile of each viewer, the content provider can then offer those viewers to advertisers with a virtual certainty that they can match advertiser and viewer together. With such rich metadata, the ability to provide advertisements that elicit positive responses by the viewer can be virtually assured. It is important to remember that John Wanamaker, the great Philadelphia retailer of the turn of the 20th Century said, "Half the money I spend on advertising is wasted; the trouble is I don't know which half."[6] That has been the traditional thinking of advertising.

With 21st Century Television, Wanamaker's phrase will never again have any meaning except as a whimsical historical comment of the ancients. With the ability to target advertising so carefully, and to understand the viewer so completely, a better phrase from the viewer's perspective will be "I don't like to watch advertisements, but I do like to watch advertisements that I like." By providing the viewer with advertisements that (s)he likes to watch, and only those advertisements, the likelihood of the advertisement being watched and watched with interest is almost virtually assured. Because an advertisement watched with interest is an advertisement that builds desire and calls for the viewer to take some action, 21ˢᵗ Century Television advertising will be delivered to an audience that is ready, willing, and eager to receive it, making the advertising extremely likely to be effective. From the advertiser's point of view, 21ˢᵗ Century Television will allow them the opportunity also to alter Wanamaker's famous phrase to "Half of my advertising may be wasted, but I know for certain it isn't my television advertising."

At this point, a question might be who would be willing to part with as much information as it would seem to take to produce a viable Aggregated Targeted Microadvertising system of advertising. According to the Pew Research Center report on "The Future of the Internet – 2010," the title itself of one portion of the larger research study says it all: "Millennials will make online sharing in networks a lifelong habit."[7] In the report, researchers from Elon University and the Pew Center's Internet & American Life Project surveyed 895 experts in a non-random survey designed to get the thoughts of those who were most likely to have informed opinions. In response to the question about how the willingness of Generation Y/Millennials[8] to share information might change as they age, 69% of the 895 expert responders agreed with the statement

> By 2020, members of Generation Y (today's "digital natives") [Generation Y has been renamed the Millennial Generation – the term Generation Y is an older term] will continue to be ambient broadcasters who disclose a great deal of personal information in order to stay connected and take advantage of social, economic, and political opportunities. Even as they mature, have families, and take on more significant responsibilities, their enthusiasm for widespread information sharing will carry forward.[9]

By contrast, only 28% felt the opposite, with 3% not responding to the question.[10]

Other more recent research studies as well as experts agree with the Pew study. Elaine B. Coleman, who was the managing director in 2012 and 2013 of Media & Emerging Technologies, for Bovitz, Inc., says, "Millennials think differently when it comes to online privacy. It's not that they don't care about it – rather they perceive social media as an exchange or economy of ideas, where sharing involves participating in smart ways. Millennials say, 'I'll give up some personal information if I get something in return.'"[11] Jeffrey I. Cole, director of the Annenberg Center for the Digital Future at the University of Southern California, is more straight-forward about the subject: "Online privacy is dead—Millennials understand that, while older users have not adapted. Millennials recognize that giving up some of their privacy online can provide benefits for them. This demonstrates a major shift in online behav-ior—there's no going back."[12] Further, in more recent Deloitte studies on television (such as the one mentioned in the section on the viewers), young viewers consistently show a willingness to surrender some or all of their privacy in exchange for information and opportunities. One really has to look no further than her/his own children's and their friends' social media pages to realize instinctively that, for young people, being social is most important, with everything else second, third, fourth, etc.

If the results of the Pew Research and other studies are cor-rect, and given the work of the Pew Research Center and USC Annenberg over the years, it is reasonable to assume they are likely to be close, then the information necessary for Aggregated Targeted Microadvertising will easily be within the grasp of content providers and advertisers through the various means described earlier.

Summary

Addressable advertising driven by aggregated targeted micro-advertising as the first step of the three-step process of ATMA,

Programmatic, and Dynamic Ad Insertion (DAI) is the future of television advertising in the 21ˢᵗ Century. As television moves more and more to the Internet through IPTV and more and more toward the notion of VOD delivery to all platforms everywhere, addressable advertising driven by ATMA in combination with programmatic and DAI, will become the de facto advertising strategy of choice. No other advertising strategy can deliver the revenues and the viewers better. Addressable advertising driven by ATMA in combination with programmatic and DAI gives the content provider and the advertiser a way to reach highly specific target audiences regardless of where they are, when they are watching, and on what platform they are watching. This ability to deliver different ads to different subsets of like viewers at the same time with a high probability of successful completion makes addressable advertising driven by ATMA in combination with programmatic and DAI an excellent partner for IPTV and VOD. The high probability of success makes each ad slot significantly more valuable. Because of the extra value, content providers can charge a premium for each ad run during that time slot. Further, the strategy makes it possible for the content provider to increase revenues by selling the same commercial slot numerous times to different advertisers targeting different viewer subsets all watching the same program regardless of time or location. Expect addressable advertising driven by ATMA, programmatic, and DAI to be the major advertising strategy for 21ˢᵗ Century Television no later than 2025, and possibly much earlier.

CHAPTER 6
PROGRAMMATIC ADVERTISING

Programmatic Advertising (or "Programmatic Media Buying;" "Programmatic," for short; or "PA") is the ability to use computer programs to determine what people watch, when they watch, where they watch, what content providers they watch, etc., and then establish "buy" opportunities on programs specifically aimed at the audiences most likely to have interests in the products of specific advertisers. Programmatic Advertising, in general, can be defined as "the use of software to purchase digital advertising, as opposed to the traditional process that involves RFPs, human negotiations, and manual insertion orders."[1] As such, Programmatic Advertising is the second step in the process of 21st Century Television advertising. It takes the information and profiles developed by ATMA and matches the profiles to programs that those people watch.

Programmatic Advertising is computer driven. It uses highly specialized computer software to determine trends in audience viewing that make it possible for agencies and buyers to purchase advertising more successfully. PA software is used in much the same way that computer software is used in brokerage houses to spot trends and allow traders to make stock trades more success-fully. The advantage of PA and its software is that it makes it pos-sible to match advertisers directly to their most likely consumers and the programs they watch. No longer is there a need to focus on ratings and the overall popularity of a program to determine where to place advertising. Programmatic Advertising is and will be used to match the advertiser with the specific programs that her/his consumers watch, regardless of the overall popularity of the program. In this way, PA makes it possible for an advertiser to expect every ad that (s)he runs to be highly successful in driv-ing the consumers most likely to purchase the product to find out more about the product and to make purchases, either through an impulse buy right on the screen or a purchase in-store or online. For the content provider, every advertising time slot becomes a "premium" slot, with the possibility (or more likely, virtual guar-antee) of selling each spot for a higher cost than a traditional ad buy would cost.

Traditionally, and still today, ratings drive the advertising world. Ratings, gathered at various times, provide the starting point for determining where to place ads to reach viewers. Ratings data include information only of the program watched, the num-ber of males who watch the program, the number of females who watch the program, and a specific set of age categories of the view-ers (for example: males 18-49, females 25-54, children 2-10). The numbers are then compared to the population of the market as a whole, and a percentage is determined. So an 8:00 p.m. program might have a "rating" of 12.5. The rating tells the programmer that 12.5% of the total market audience watched the program, or one in every eight people (in this example).The same numbers can also be compared to the number of people in the market who actually had their TV sets turned on to any program at that time. Taking the same example, the program might have a "share" of 25. The share says that 25% of people who had their TV sets on at 8:00 p.m. watched the program.

Based on the rating and the share of a program, the content provider will set a price for an ad, depending on the time length of the ad. The larger the rating and the share, the more the content provider can charge for the ad. Notice, the only information the ratings data provide are the overall number of the viewers watching the program, and how the numbers data break down among males and females and a handful of age categories. On that information alone, advertisers have been placing their "shotgun approach" advertising on content providers' programming, hoping to successfully reach 15-20% of the program's audience. The traditional mantra has been that you "sell on your quantity or on your quality." That means that if the ratings are large, boast about your overall rating and share. If the ratings are small, then find what audience is your largest and sell that group to advertisers.

Throughout the history of television (and radio before it)

- programs have stayed on the air or been cancelled, depending on how the ratings were for that program over time. Today, that time frame can be as short as just a few weeks, given that weekly ratings of all programs aired each week are available to the content providers;
- program schedules have been changed, including moving successful programs to other days and time slots to bolster or serve as an anchor program for a weak ratings night (almost always to the detriment of that successful program), and
- advertisers have paid billions in money to purchase air time on television programs, with little chance for large rewards, given the shotgun approach to television advertising. However, when the shotgun approach is the only option, advertisers learn to live with the results and find ways to justify their actions.

Today that seems a little silly, doesn't it? That's why PA is so important. More specifically, programmatic television advertising can be

defined as "the data-driven automation of audience-based advertising transactions. It inverts the industry standard, in which marketers rely on show ratings to determine desirable audiences for
their ads. Instead, with programmatic tech, marketers use audience data to pipe advertising to optimal places."[2] Eric Blattburg
writes that programmatic advertising provides more specificity. He
says, "Rather than relying on ratings for specific shows or channels, marketers can use programmatic tech to reach a more specific subset of consumers, like men with a $50,000 income who
own an Android device."[3] Blattburg goes on to say, "They [market-
ers] don't care if that ad shows up on *X Factor* or *X Games,* as long
as the target audience is watching."[4] Sound familiar?

Let's break this down more: Looking back, the definition says
PA is "data-driven automation of audience-based advertising transactions." Let's look at the individual parts of that first sentence:

- Data-driven – with PA there is so much more information
  to consider than just the ratings data. All the data gathered
  through ATMA can be used to make decisions in PA. ATMA
  provides the data profiles of viewers and PA uses that data
  to determine the best matches in terms of programs.
- Automation – the decisions are automated; all the data are
  loaded into the software package and it does the matching.
- Audience-based advertising transactions – just like with
  ATMA, note that the unit of focus is the viewer, not the
  program. Traditionally, the focus has been on the program
  and matching the ad to the viewer through the program.
  PA changes the dynamic.
- "It inverts the industry standard." This is what is so hard for
  traditionalists. PA changes the whole way advertisers and
  content providers have been doing business. Instead of relying on "show ratings to determine desirable audiences for
  their ads," the determination is made through ATMA data,
  of which show ratings are a miniscule part.

How Programmatic Advertising Works with ATMA and DAI

As mentioned at the beginning of this chapter, Programmatic Advertising is the second step in the addressable advertising sequence. This sequence of Aggregated Targeted Microadvertising, Programmatic Advertising, and Dynamic Ad Insertion is designed to work together to provide the best opportunity possible to reach the viewer no matter where or what program (s)he is watching. Here's the way programmatic advertising works with DAI driven by ATMA:

ATMA first determines the specific audience for a product or brand and makes it possible for the advertising creative department to design the ad campaign to reach that specific audience as has been described in the previous chapter. Programmatic technology, then, uses the data provided by ATMA to determine where those audiences are – the program(s), the time(s) of day, the location(s),[5] content provider(s),[6] etc. – and establishes "buy" opportunities to reach those ATMA audiences. In other words, Programmatic Advertising matches the audience information and the advertising creatives with the programs that are best suited to those audiences and creatives and purchases advertising time in those programs. This matching of the programs with the audiences and creatives ensures a high probability of success for those advertising vehicles with the audiences.

Once ATMA and Programmatic have done their jobs, the Dynamic Ad Insertion technology takes over, inserting the specific advertisements determined by the previous two steps of the process into those locations where the audiences for that brand are located, and reaching the audiences that give the advertisements for the brand or product the highest possibility of success.

Programmatic Advertising is not the least interested in **how many** people watch a specific television program. It is only slightly more interested in the numbers of males and females, and what different age groups, are watching program, as those are just two of the many variables ATMA will consider in developing the profile of the targeted consumer. Instead, Programmatic is interested in which groups of individuals with those very specific characteristics choose a particular program and then matching those groups with their specific advertising choices to make a synergistic whole. Programmatic Advertising technology, once again then, uses the data provided by ATMA to determine where those audiences are – the program(s), the time(s) of day, the location(s), content provider(s), etc. – to establish the most successful "buy" opportunities to reach those ATMA audiences.

10 Things You Should Know About Programmatic Advertising
In June 2015, Alex Kantrowitz published the article "10 Things You Need to Know about Programmatic Buying" in the web edition of *Ad Age*.[7] That article – which, admittedly is about more than just television advertising – lays out guidelines for consideration for anyone who is (was) considering becoming involved in using Programmatic Advertising. As he puts it in his article, "When surfing the wave of automation, here are 10 things you should know:" (Author's Note: Because of the 2015 date of the article, where possible, updated information has been included simply to make the points more relevant. Those specific updates will have separate citations to denote their inclusion. Otherwise, all information will come from the aforementioned article.)

1. **The pie is growing (fast).**
According to Zenith's Programmatic Marketing Forecasts report, 65% of the money spent on advertising in digital media (of all kinds) in 2019 will be handled through programmatic advertising. That 65% represents an ad spend of some $84 billion that will be spent using Programmatic Advertising, an increase of $14 billion over the $70 billion spent in 2018. Further, the Zenith report forecasts that in 2020 the ad spend using programmatic will again increase to $98 billion, or 68% of total digital ad

spend. In other words, by 2020, two out of every three dollars spent on digital advertising will be spent using Programmatic Advertising.[8]

2.   It's not just for direct response anymore.

Programmatic-buying systems today are being used by legacy television outlets, both the networks and the local stations, as well as virtually all of today's "over the top" TV channels. For instance, PremiumMedia360's automated TV advertising platform is backed by some of the major players in local television, including group owners Hearst Television, Nexstar Media Group, Raycom Media, Tegna, Tribune Media, and the powerhouse giant Sinclair Broadcast Group. It's available to 320 TV stations nationwide, to go along with competitors WideOrbit (WO Programmatic), Cox Media Group's Videa, ITN Network (ProVantageX), as well as others. Of those mentioned WideOrbit offers programmatic ad inventory across more than 1,000 YV channels, Videa is in 150 markets that reach 80.2 million households, andProVantageX is on more than 800 TV stations.[9,10]

3.   Data rules.

One of the great aspects not mentioned before is that, with programmatic advertising platforms, ad buys can change quickly as needs change in response to the changing audience for any program. The data results from the original ATMA platform will be continually evolving and changing as audiences evolve and change for each program. Those evolving data can then be used by the programmatic ad platform to evaluate what's working best throughout the campaign in terms of the programs for the advertiser—which geographies, times of day, audience segments, publishers – and can further refine and target ever more effectively the campaign buys. In doing so, the advertiser is paying only for highly effective ads. This is a radical change from traditional ad buying, where a buyer agrees to run a certain number of ads with a publisher and is locked in to the contract.

4. Brands are taking it in-house.

Today, more and more advertisers and their agencies are developing their own programmatic advertising platforms. In an October 2018 article in *eMarketer*, Ross Benes writes that, in an Association of National Advertisers survey, 78% of the respondents (US client-side marketers) reported having an in-house agency that was capable of programmatic advertising. Further, the survey found that – for almost 40% of the respondents – cost efficiencies were the reason for taking the programmatic ad buying platform in-house.[11] In addition, an IAB report reveals that approximately 40% of advertisers are executing programmatic in-house programmatic trading.[12] While programmatic advertising mentioned in the surveys above is not specific to television, the increasing use of programmatic advertising over time is evidence that programmatic is essential and is growing.

5. Mobile is a major issue.

While mobile was a major issue when these ten considerations were originally written, the fact is that, for 21st Century Television, mobile will be of minor concern, given that the behavioral targeting capabilities of programmatic systems depend less on the platform and more on the individual(s) that are targeted. Whether a viewer or an entire subset of the viewing audience being targeted is watching the program on a television set, a computer, a tablet, or a smartphone, the advertising will be the same. It won't be necessary to change the products just because of the platform, although the technical aspects of the creative will likely have to be altered to fit the platform used for watching. As such, while for other forms of digital advertising mobile may be an obstacle to overcome, for 21st Century Television, addressable advertising on the platform won't be an obstacle. At worst, it will be a minor nuisance.

6. Social networks are gaining clout.

Once again, like #5 above, this item is stated more for a 2015 audience than today's audience. Yes, social networks have clout, but, as will be discussed in detail in the chapter on promotion, social networks will play a complementary role for 21st Century Television, being more of a collaborator than a type of competitor. Certainly, over the past few years, Facebook, Twitter and LinkedIn have gobbled up programmatic ad-tech companies. They have been on the forefront of programmatic advertising, making excellent use of the system to sell ads across the web, not just on their own platforms. Further, these social-media networks have reams of login data, allowing them to connect user identities across devices. However, those data can also be used by content providers and advertisers as well, making the collaboration of social media network, ATMA, programmatic, and DAI an even more powerful and lucrative way to reach consumers with the products, services, and messages they want to hear and respond to in a positive and highly effective manner.

7. Fraud is still a problem.

Fraud continues to plague advertisers and content providers, although less so than for other forms of digital programmatic advertising. While the programmatic ecosystem is especially susceptible to a fraudulent practice called "URL masking," the bigger concern is the impact of "bots" on ad viewership. The concern is that these bots are "fraudulent viewers"; they appear to be a viewer when they are not. The inclusion of a bot in the results makes an audience seem larger than it is, necessitating refunds when they are discovered. However, with ATMA as the beginning step in the addressable advertising sequence, a bot is much less of a concern, because the procedures and the requirements of ATMA make it extremely difficult, if not impossible, for results to be skewed by bots. Therefore, advertisers can feel much more assured that when ad buys are made, they are reaching their intended audiences in the correct numbers that are anticipated. Bots will no longer be able to influence the numbers because they will simply not fit into viewer profiles.

8. **The big players are set.**

For 21st Century Television, the above statement is not correct, although it is more correct for overall digital advertising. There are still big, new programmatic technology companies coming into the ecosystem because addressable advertising is in a much earlier stage of development than other forms of digital programmatic advertising. As recently as last year (2018), new companies working in addressable television advertising were opening their doors as the 21st Century Television universe continues to grow and expand and more content providers embrace that universe.

9. **It's not just banner ads.**

For 21st Century Television, banner ads will play only a miniscule role in generating revenues as addressable advertising becomes the advertising vehicle for content providers and advertisers alike.

10. **It can be good for viewability.**

Yes, it can and it will

Programmatic Advertising, as mentioned earlier, is the second step in the three-step process of advertising of 21st Century Television. It is the future of how advertisers will be matched with their customers through TV programs.

Artificial Intelligence, Machine Learning, Deep Learning, and Programmatic Advertising[13]

Innovative technologies such as machine learning, deep learning and programmatic advertising are some of the hottest topics in advertising right now. All these terms are often used interchangeably, but they are definitely not the same thing. All the buzz around these technologies comes from the fact they are transforming the way industries work and bringing significant changes and improvements to marketing and social media advertising.

What, then, makes machine learning, deep learning and artificial intelligence different? How is programmatic buying related to all of this?

Programmatic Advertising

Programmatic ad buying is the process of using software to buy digital ads. Instead of going through human negotiations and manually inserting orders for ads on digital platforms like Google and Facebook, media buyers use software to go through an auction-based process to have their ads served in a network of their choice.

Buyers use big data to target their most valuable customers by segmenting their audience through characteristics like age, gender, region, among others. They go, then, through an auction process and choose if they will pay the price an ad is worth at the moment to have their ad seen by this specific audience.

Programmatic ad buying reduces costs and improves brand performance in the ad industry. This technology is now getting an upgrade and advertising software is becoming smarter due to artificial intelligence.

Artificial Intelligence

Artificial intelligence is the concept of replicating human intelligence in machines so they can perform activities that would require the human brain [to be] involved, such as making data-based decisions.

AI-powered systems help companies save money by working at a faster speed than humans beings and with less potential errors. If you apply this technology to the advertising industry, you bring efficiency to the media buying process, freeing media buyers from tedious work to focus on strategic and creative tasks.

Machine Learning

Machine learning is a type of artificial intelligence that provides computers with the ability to learn things by being programmed specifically to certain tasks, improving its knowledge over time the same way the human brain does.

It focuses on mimicking our own decision-making methods by training a machine to use data to learn more about how to perform a task. Imagine you drive to work every day. Over time, after trying different ways to arrive at your destination, you will learn which path is faster or maybe which route is better according to the day of the week.

This is how machine learning works. You feed a machine with large amounts of data so it will analyze information from the past and learn from it to apply the newfound knowledge to any new data it receives in the future.

In terms of advertising, machine learning algorithms can analyze data and draw conclusions from it. It means you can basically replicate the brain of an experienced buyer in a computer so it becomes capable of diagnosing, predicting and planning things.

Deep Learning

Deep learning is a branch of machine learning. It also generates insights based on data but focuses on more specific aspects of machine learning. With deep learning, you combine multiple layers of information to go even deeper in a specific subject.

In advertising, you can use deep learning as one way of combining different factors to draw conclusions from them and learn, for example, how people who live in a particular region, have a specific age and like folk music are more likely to buy sports clothing.

Get ahead of the competition

When it comes to computer intelligence, some of the terminologies can get a bit muddled by confusing or incorrect definitions. By understanding what these terms mean and how these technologies affect your work you can potentially get ahead of the competition.

How Does AI Tie To Programmatic Advertising?[14]

Programmatic advertising is the automated process of buying and selling ad inventory through an exchange, connecting advertisers to publishers. This process uses AI and real-time bidding for inventory across all channels. AI can analyze the complexity of media buying via programmatic in a way that is not humanly possible or not possible by human media buyers and planners.

Artificial intelligence gives marketers the ability to take control of their data to achieve the results they are looking for. It gives them the power to optimize their digital marketing campaigns pre-bid. They can determine exactly where they should place their bid, how much they should bid, and when they should bid — this is all before they even place the bid. The goal is to help them monetize their campaigns and deliver the best possible return on investment.

The purpose of AI is to take the massive amount of consumer data that's collected, and analyze the information it contains about consumer demographics, interests, and purchasing preferences. Marketers then use this analysis to determine the right audience for an ad so they can create more focused and targeted ads, which leads to better campaign results. This is especially helpful in video ad campaigns, where the proper placement and timing of the ad is a critical aspect of a campaign's success.

AI and predictive modeling techniques boost campaign effectiveness by accelerating the decision-making process in terms of determining what ad should be delivered to which user, what type of format should be used, and the best time to deliver it.

Benefits and Challenges

In terms of benefits, AI can increase marketers' knowledge and understanding of consumer behavior in ways that were never possible before, making for more relevant, cost-effective, and optimal advertising. AI can process and analyze the quantity and complexity of big data in a way that hasn't been possible before. The result is marketing on a scale that's never been imagined, let alone achieved.

AI delivers better ad campaign control because it employs more quantified and automated strategies. These strategies can be used to improve the customer's shopping experience, to test advertising campaigns, and to make campaign bidding decisions more cost-efficient. The main **challenge** with AI is having the technology predict the right action, in context. Algorithms need to work seamlessly to make decisions in real time. Not every organization has the resources to build these capabilities. Getting from the theoretical to reality is a challenge that not all companies are prepared for. It's also an issue of personnel. The motivation for a company to become fully automated is often questioned by workers who fear what that might mean for their job security.

Machine Learning's Role

Machine learning algorithms operate 24/7. The learning never stops, enabling greater awareness and knowledge of even the most subtle changes in market behavior, which can be used to make better marketing decisions. It's a combination of developments, including more powerful computing, big data, and advances in deep learning technology. This combination has made it possible to create and maintain large datasets that deep learning algorithms can analyze for marketing purposes, such as identifying trends and making predictions. For example, with AI, large amounts of data, such as browsing or shopping histories, can be analyzed to tell advertisers where, when, and whom to target.

AI is always learning. The more it's used, the smarter and more efficient it becomes. So AI learns which users to target and which users not to target, which users are likely to engage, and which users aren't likely to engage. From this, AI can determine only the most relevant users to target, which means fewer wasted ad impressions and highly defined, targeted ad campaigns.

Summary

Programmatic Advertising, as mentioned earlier, is the second step in the three-step process of advertising on 21ˢᵗ Century Television. PA will be driven by AI, machine learning, and deep learning to create deep and abiding connections between advertisers, content providers, and viewers. It is the future of how advertisers will be matched with their customers through 21ˢᵗ Century Television.

CHAPTER 7
DYNAMIC AD INSERTION

Dynamic Ad Insertion, or DAI, is the third step in the process for delivering 21st Century Television advertising. DAI can be defined as the ability to simultaneously insert different ads into different video streams, for different viewers. DAI makes it possible to send a variety of different ads to different viewers simultaneously using different streams. It is the end result of addressable advertising.

History of Dynamic Ad Insertion

Of the three steps in the process, DAI is the oldest. The forerunner of DAI goes back to the 1930s when magazines would use "split-run" advertising for research. A magazine would choose a few test markets and run an ad for a product to half the subscribers, and a different ad to the other half. The more popular ad would then be used nationwide.

Post-1975, when cable had begun to flourish as an alternative to OTA television, cable companies also began using split-run advertising on a limited basis, again for research purposes. The process would be the same – half the subscribers got one ad, the other half got a different ad for a product. The more popular one would then be used nationwide. In both cases, the use of the split-run was for research purposes, not for directing advertising toward different audiences.

The first real discussions and attempts to use DAI came in the mid-to-late 1990s, when radio stations first began streaming their signals over the Web. Numerous small start-up companies specializing in various forms of DAI – some simple and others more complex – began to make DAI available to the terrestrial radio stations and the Internet-only stations that were also getting underway. For those start-ups, the world came crashing down when, in the U.S., the Federal Communications Commission (FCC), responding to the pressure of the music industry, ruled that each stream was a separate program and could be charged copyright royalties. So, if a station had 50,000 listeners to its streaming program, according to the FCC, the music industry could demand payment as if the station was broadcasting 50,000 different programs simultaneously.

The costs were more than stations could afford, so stations shut down their streaming services, making DAI advertising useless. Every early DAI company went out of business.

Dynamic Ad Insertion Today

Today, the DAI industry has recovered from the situation of the 1990s, but has not yet truly flourished. Only in the past few years have content providers begun streaming their programming to audiences, and then only on a delayed basis. Further, the legacy media companies have been reticent to move to streaming significant portions of their programming in real time, fearing that to do so might reduce the viewership in traditional television. Additionally, in the U.S., the content providers receive billions of dollars each year from the cable and satellite companies in fees for the right to carry their programming – money which would go away with the move to all streaming. Also, there is still the belief – mentioned earlier – that people will not watch commercials if they watch on the Internet. And, the content providers feel that it is better to keep what they have instead of trying something new. As such, the DAI industry is still in its early stages, waiting for the time when the content providers finally move to IPTV and their industry begins to flourish.

Where DAI is and will be most effective is with video-on-demand (VOD), which, of course, is the way 21st Century Television will be delivered. That fact suggests a dynamic future for DAI.

Video-on-Demand and Dynamic Ad Insertion

Video-on-Demand has become a popular platform not only for consumers but also among television networks, advertisers and cable service providers for its ability to be measured and therefore monetized. It has created the atmosphere for new advertising opportunities in the non-traditional television world, including Dynamic Ad Insertion. DAI expands advanced advertising opportunities by allowing cable providers to target ads that can be swapped in and out of VOD content.

Ads can be 15, 30 or 60 seconds in length. The 30 second ads are the most popular in the VOD platform. Ads can be inserted into VOD content through pre-roll, mid-roll and post-roll formats. Dynamic Ad Insertion has created opportunities for all parties involved to gain more value out of this form of advertising.[1]

(Technology providers like Canoe Ventures, which powers VOD ad insertion for major broadcast and cable networks, have now expanded their VOD DAI platform across 130 DMAs and have already begun to see an increase in the amount of ad insertions. In the 2014 year-end report issued by Canoe Ventures, viewed ad impressions increased by 215% from the first to the fourth quarter of 2014.[2])

One of the major advantages for VOD advertising is that most ads are non-skippable by the viewer; however, that function is determined by individual network programming providers. One VP of Research & Consumer Insight recently noted, "After a few technological missteps early in the process, DAI/ VOD is coming into its own and is now offering advertisers and agencies another viable option to reach target consumers using video ads placed within a high-profile, premium programming environment."[3]

Note that, because different ads are placed simultaneously into different streams of the same program to reach the different targeted groups, several results occur:[4]

1. Each individual ad reaches a smaller group of people than ads today.
2. By targeting the viewer with an ad for the product (s)he wants in the way that (s)he likes, the ad has a higher success rate – somewhere around an 80% or greater success rate.
3. Because the product is something the viewer is interested in, (s)he will almost always watch the complete commercial, making the high success rate possible.
4. Because of the high success rate, content providers can charge a "premium" on each ad (each ad is more expensive).
5. Everyone gets what (s)he wants.

Let's look at each of these results individually:

1. Each individual ad reaches a smaller group of people than ads today.

Addressable advertising is used to segment the audience. The first step is ATMA; the second, Programmatic. Once the first two steps have been completed, the audience for a given show and its accompanying ad time has been segmented and the appropriate advertising message has been placed to reach every segment of the audience, according to the decisions made by steps one and two. As such, because the audience has been segmented, and each segment is receiving a different commercial for a different product or in a different way from the other segments, each segment is smaller than the whole audience watching the various commercials during any given commercial time slot.

2.   By targeting the viewer with an ad for the product (s)he wants in the way that (s)he likes, the ad has a higher success rate – somewhere around an 80% or greater success rate.

Today's television markets' ad time is designed to reach the entire audience during each commercial time slot. While not totally undifferentiated, given that ratings do provide some measure of the viewership, an ad that works is one that successfully reaches 15-20% of the viewing audience. By segmenting the audiences through the in-depth analytics of ATMA, each member of the audience only sees ads specifically designed and chosen for her/him every time there is a commercial time slot that's filled. By providing each audience member with ads for the products and services of most interest to him/her, the likelihood of a successful ad increases dramatically. Added to the specific ad for the specific viewer is the opportunity for the viewer to immediately click on the product during the ad and make an immediate impulse buy, thus adding to the success of the commercial.

3.   Because the product is something the viewer is interested in, (s)he will almost always watch the complete commercial, making the high success rate possible.

As has been pointed out, viewers hate commercials except for the commercials they like. The three-step process of ATMA, Programmatic, and DAI gives the viewer those commercials that (s)he likes, in the way (s)he likes them, with the creatives that (s)he likes, on the platforms that (s)he likes, in the programs (s)he likes, and when (s)he likes. If the viewer is getting all those things that (s)he likes, the viewer is virtually guaranteed both to watch the commercial and to interact with the commercial in a positive way. In fact, when DAI first begins providing the commercials for the viewers that ATMA and Programmatic have planned, the viewers will most probably be so happy and amazed that it appears someone is actually listening to them that the success rates of the ads will be close to 100%.

4. Because of the high success rate, content providers can charge a "premium" for each ad (each ad is more expensive).
Granted, each ad that is delivered to each of the various segments of the viewing audience for a program will be smaller – possibly significantly smaller – than the audience for today's "shotgun blast" commercials. However, for the advertiser, that will mean her/his ad is no longer falling on "deaf ears," those people for whom the ad has no relevance or cannot reach for whatever reason. Instead, the advertiser will be reaching only those viewers who are his/her most likely customers or potential customers. To be able to reach those customers and not have to waste money on all the non-customers like TV advertising today, the advertiser will happily pay a premium for the chance to reach her/his customers with such a highly targeted opportunity. While any one commercial for any one advertiser will not equal the full price of today's "shotgun" commercial, the ability for the content provider to stream multiple commercials to multiple segments of the audience simultaneously during each commercial slot, will far exceed the price for the one "shotgun" ad that airs today.

As a simplified example, assume the audience for a particular program is 100 people. By today's standard one ad is run during a commercial time slot, costing $1000. For that ad to be considered successful today, the ad must successfully reach 20 people with the other 80 ignoring the ad. The cost, then, per successful viewer is $50 (the rest do not count because they are not interested). Under the addressable advertising three-step process of ATMA, Programmatic, and DAI, it's determined that there are five different segments of 20 viewers each out of that 100 total viewers. Further, for an ad that reaches the usual 20% of each group of 20 in the five segments, the cost for each of the five ads reaching each of the five segments is $20 per viewer (because the total number of viewers reached is lower). Under the current pricing method, the total revenue would only be $500, or a loss of $500.

However, because the successful commercial under the addressable advertising process reaches, at a bare minimum, 80% of the viewers, there is a premium to be added to the price of each of the commercials that go to the five segments of 20. Let's say the premium is $5.00 per viewer for every increase of 20% in the number of viewers – a reasonable sum. Now, the price per viewer goes from $20 per viewer to $35 per viewer (the original $20 plus $5.00 for each of the three 20% increments between 20% and 80%), and the total number of viewers for each ad has gone from 4 (20% 0f 20 viewers) to 16 (80% of 20 viewers). To reach those 16 viewers out of the 20, the ad costs $560 and the total revenue for all five ads run during that commercial time slot is $2800, or 2.8 times the cost of the original ad. Notice also, that for four times the number of successful connections – 80 out of 100 – the total cost went up only $1800 which was spread across all five advertisers. Furthermore, all five advertisers got a much higher success rate.

5.   **Everyone gets what (s)he wants.**
The advertiser knows the ad will be successful because (s)he can expect a success rate of at least 80%. The content provider makes a lot more money because (s)he can sell each advertising spot at a premium and sell multiple spots for each commercial time slot, the number of different ads being dependent on the number of

different segments of viewers there are for the program. Finally, the viewer is happy with the ad (s)he is watching because it is for a product or service (s)he wants or uses, in the way (s)he likes them, with the creatives that (s)he likes, on the platforms that (s)he likes, in the programs (s)he likes, and when (s)he likes. The result of the three-step process of ATMA, Programmatic, and DAI is truly a "win-win-win" situation.

Benefits of DAI driven by ATMA assisted by PA

Combining DAI with ATMA and Programmatic produces a number of benefits for the content provider and the advertiser:[5]

1. Ads in streams cannot be skipped by the viewer
2. The ads can be used throughout all the viewers' choices. Remember, it doesn't matter the program, it's the individual that is important.
3. Ads can be completely interactive.
4. Ads can be delivered to their targeted audiences wherever they are worldwide.
5. Different ads can be delivered to different audiences at the same time.
6. Ads can be placed in every program in the content provider's library of programs. This ability makes every program valuable to the content provider and the advertiser.

As with the results, let's look at each of the benefits in depth.

1. Ads in streams cannot be skipped by the viewer

Today's advertising on television allows the viewer numerous ways to ignore the commercials. Using DAI, the fact that the ad cannot be skipped requires the viewer to watch the ad (or ignore) at real time speed. The viewer can still leave the room or multitask with other items, but using the three-step process of addressable advertising will make the advertising so enticing that the desire to ignore the ad or skip it will be eliminated, or, at worst, reduced significantly. When the viewer is offered an ad that is for a preferred product or service, delivered with the right creatives, in the manner in which the viewer prefers, on the platform of the viewer's choosing, the viewer is going to watch and enjoy the ad. Further, by making it possible for the viewer to click on and immediately go to the website of the product virtually assures the advertiser of numerous impulse buys.

2. The ads can be used throughout all the viewers' choices. Remember, it doesn't matter the program, it's the individual that is important.

In 21ˢᵗ Century Television, the viewer is in charge. (S)he selects the program, the time of watching, the location for watching, the platform for watching – the viewer is truly in charge. By being in charge, the viewer makes it clear as to which programs (s)he prefers and chooses to watch, and the content provider as well as the advertiser knows this information. Most likely, as today, viewers will compile a list of their favorite programs they want to watch during a certain time period (day, week, etc.) on their connected, smart television sets. Further, they may also set time schedules of when they have the time and prefer to watch. That information will also be on their television sets and will be known to the content provider and the advertiser. (The author uses the television set as an example, but the schedule could just as easily be on a computer, tablet, or smartphone. Regardless, the information will have to be shared with the content provider and the platform connected to the provider when watching the chosen show.) Regardless of what the program is or when, where, and what platform it is watched on, the advertising will be available because the ad inventory will be keyed to the individual viewer through the repeated demonstration and choices made by the viewer. Regardless of what program the viewer chooses to watch, the correct ad will be sent to her/him for viewing and enjoyment.

3. Ads can be completely interactive.

Because IPTV is interactive, the ads also can be interactive. While the interactive nature is discussed more fully in the following chapter on product placement, it does deserve mention here as well. Ads on 21st Century Television (as well as programs for that matter) will be fully interactive. The viewer is also in charge when it comes to responding to the ad and to the product or service. When an ad is presented to the viewer, (s)he can choose to click on the ad at any time and be taken directly to the home page of the website, or even to the "buy" page if the viewer so chooses. If the viewer goes to the home page, then (s)he will have the opportunity to see what choices there are for the product, including such things as size, color, levels of choice (e.g., basic, deluxe, premium), warranty information, and price(s), just to name a few of the options that could be available. Remember, as discussed earlier, television sets will continue to grow in size, smartphone screens will likewise, and all platforms will have the ability to display multiple pages across a single screen. The viewer can continue watching the ad (or the program if the ad is over) on one portion of the screen while making his/her decision about the product on a separate portion of the screen.

4. Ads can be delivered to their targeted audiences wherever they are worldwide.

As will be discussed in Chapter 10, 21st Century Television is a complete global commodity. It will be able to reach audiences no matter where they are in the world. As such, content providers and advertisers will find new audiences and potential customers in other countries to reach with their programming and their advertisements. For many companies, 21st Century Television will be a way of allowing companies opportunities to establish new markets in countries where previously they had not been able to or could not afford the cost to establish the market. The global nature of

21ˢᵗ Century Television will also make it possible for viewers to watch their favorite programs and advertisements whenever they go abroad. Regardless of the time zone, the time of day, or the country they're in, viewers will be able to watch their favorite programs and advertisements in their own languages when they want, where they want, how they want, and on whatever platform they want or is available. Further, viewers will be able to enjoy the ability to order whatever product they want or need immediately and have it shipped directly to their locations wherever they may be. All that will be needed is an Internet connection, which is possible virtually everywhere throughout the world.

5. Different ads can be delivered to different audiences at the same time.

The great thing about DAI is that the ability to insert advertisements into a program delivered by IPTV is limited only by the number of streams. Because there will likely be more than one subset of an audience for a program, each commercial time slot can be sold numerous times to reach each of the different subsets. As discussed earlier, the ability to reach these subsets with the ads they want for the products and services they want, in the way that they want, means that each ad will have an extremely high rate of success. Content providers, then, will be able to sell those audience subsets to the appropriate advertisers for a premium, and therefore make even more money than they are capable of making using today's advertising strategies and techniques (see the described example from earlier in this chapter).

6. Ads can be placed in every program in the content provider's library of programs. This ability makes every program valuable to the content provider and the advertiser.

Because every program now has value no matter how new or how old the program is, there will be an audience (and likely more than one audience subset) that can be reached by 21st Century Television addressable advertising. Now, every program has value and every viewer has value, because even small audiences can produce revenues for both the content provider and the advertiser that have not been available before, given the nature of today's television. No longer will shows have to be cancelled; in fact, cancellation will be a thing of the past. Loyal viewers of any program, even ones that would normally be cancelled in today's television scheme, will continue to watch their favorite programs, potentially again and again and again, receiving and appreciating the commercials from advertisers they prefer, who remain dedicated to the program and the small, but extremely loyal, audience. That appreciation will produce small but continuing revenues from those audiences which will likely be an important supplement of revenue for both the advertiser and the content provider.

Summary

Dynamic Ad Insertion is a concept that has been discussed in the TV industry for years. Since the earliest days of streaming, first with radio and then with television/video, DAI as a source of revenue has been a topic of continuing interest and discussion. The reason is simple: DAI, along with ATMA and Programmatic – the other two steps in the addressable advertising process – has the potential to create the largest single revenue opportunity for content providers and advertisers in the history of the electronic media, and, when fully implemented, will provide the monetization of a coming "diamond age of television" the likes of which the world has never seen..

DAI is the final step in the financial engine for 21st Century Television, as it makes it possible for advertisers to present fresh, relevant, highly targeted ads to viewers/customers anywhere, anytime, anyway, and on any platform, as well as reaching audiences around the world, opening new markets to advertisers and viewers/

consumers to new products and services. In this way, content providers can add billions (potentially even trillions) of dollars to their bottom lines.

To conclude – the three-step process of dynamic ad insertion driven by aggregated targeted microadvertising in combination with programmatic advertising is the future of television advertising in the 21st Century. As television moves more and more to the Internet through IPTV and more and more toward the notion of VOD delivery to all platforms everywhere, the three-step process of DAI driven by ATMA in combination with programmatic, will become the de facto advertising strategy of choice. No other advertising strategy can deliver the revenues and the viewers better. It is truly a win-win-win situation for all concerned and the key to 21st Century Television's diamond age.

Section 2:
Other Monetization Opportunities

CHAPTER 8
PRODUCT PLACEMENT

Product placement is "an advertising technique used by companies to subtly promote their products through a non-traditional advertising technique, usually through appearances in film, television, or other media."[1] While you might question the "subtlety" of product placement these days, product placement is an ever-growing way of advertising on today's television and will be an even more integral part of 21st Century Television. A simpler way to define product placement is – if you see a product being used in a production or is seen in a production and you can read the name or recognize the logo or product. As an example: You are watching a television program and see a Samsung computer in a scene and you can recognize it or see the Samsung name or logo, that's product placement.

While product placement has traditionally been considered a part of the movie industry, product placement on U.S. television will reach $11.44 billion by the end of 2019, according to PQ Media research.[2] While legacy television still accounts for more than 70% of all product placement dollars, OTT television (in its broadest form) is the fastest growing, accounting for more than 25% of ad dollars spent in 2017. This percentage includes the original programs that are on both Netflix and Amazon Prime Instant Video, neither of which accepts traditional advertising commercials.

History and Background of Product Placement

While the history of product placement on television is a relatively short one, product placement has been used extensively in movies from the earliest days. One of the earliest examples of product placement came in the 1919 movie *The Garage,* directed by and co-starring Fatty Arbuckle, when Red Crown gasoline was displayed during the movie.[3] The use of product placement in silent films established the ability of movie production companies to finance in part or in whole their movies by allowing companies to place their brands in prominent locations during a movie so they could be easily seen. Product placement in movies continues on a regular basis to this day.[4]

Product placement in television has also existed since the early days, often overtly in children's television. Though the practice is not called product placement, children's programming has always had a product placement aspect to it, in that the associated toys, clothing, cereals, etc., have always been tied to the programs starring the cartoon figures or children's shows actors and actresses. Going back to the earliest television days (of my own youth, I might add), children could buy – or have their parents buy for them – Woody Woodpecker stuffed toys, Superman clothing, Roadrunner wind-up toys – pretty much something for just about any television show that children watched on Saturday morning or after school. The characters themselves became, in a real sense, the embodiment of product placement. In fact, Saturday morning cartoons became to be so much of a thirty minute or hour-long product placement advertisement (the *Smurfs* is generally cited as a prime example of such programming during this time) that the FCC finally acted to limit the connections between the programs and the adjacent commercials that could be shown.[5] Other countries around the world also limit such forms of advertising.[6]

Adult television programs have also had their share of product placement, although there was more subtlety to them in the earlier days. Automobiles, because of the necessity of the characters to move from one place to another, figured heavily in product placement on television. Crockett and Tubbs chased criminals in their Ferrari Testarossa in *Miami Vice; Magnum P.I.'s* Tom Selleck drove a Ferrari 308i. For the Duke boys in *Dukes of Hazzard*, it was a Dodge Charger named "General Lee." Even early television programs such as *The Prisoner,* made good use of automobiles. In the case of *The Prisoner,* the automobile was a Lotus 8.[7]

Game shows such as *The Price Is Right, Let's Make A Deal,* and today's *Wheel of Fortune* all make use of extensive product placement. *The Price Is Right* was virtually an hour-long product placement as the viewers looked forward to hearing Johnny Olson's call, "It's a new car!" in his own inimitable way. Game shows are adept at product placement, swapping extensive publicity and focus on the product in exchange for having the product given to the program for the publicity.

Today, of course, we see product placement used extensively in television programs. Whether it is Jack Bauer driving around in a Ford Expedition in the program *24* or it's *The Biggest Loser* contestants making a visit to Planet Fitness (or Anytime Fitness, or the gym of choice for the season) or filling their water bottles from a Brita pitcher, television programs today find that product placement is an exceptional form of additional advertising revenue to help fund the production and distribution of the program. Reality shows, especially in the U.S., are just one product placement after another. Often, the product placement is very obvious. Everywhere a reality show contestant goes, everything (s)he does, everything (s)he wears is a product placement. If the contestant goes to a luxurious spa for a weekend – it's a product placement. If the contestant rides in a limousine or flies in a private jet – that's product placement.

So product placement is in virtually every television program there is, including live sporting events. Professional golfers are walking product placement billboards – in addition to getting paid for using various companies' products, they are also paid every time they appear on camera – a real additional incentive to play well each week. Professional race car drivers are also walking billboards as well as their cars; they have similar arrangements with the companies for appearing on television. Think about football teams, especially European futbol teams (it's what, in the U.S., is called soccer), and maybe Asian teams as well. Each team player has the name of the team's sponsor – some corporation – sewn in large, bold letters across the front of the shirt. The team name and crest is a small, often unrecognizable to anyone but ardent fans, patch over the left breast of the player. It's obvious the importance of product placement in the sport. In Korea, major league baseball teams are not known by the city where they are located like in the U.S. Rather, they are sponsored by major corporations, so every player is a continuous product placement for the team and even the mention of the team name (e.g. Kia Tigers located in Gwangju or LG Twins located in Seoul) is a product placement. When teams are on television, the entire game – from pre-game through to post-game – is a product placement for the two teams and the corporations that sponsor them.[8]

As such, advertisers are happy to associate their products directly with a popular program – or even a less popular one, if the program has the right demographic makeup – because viewers cannot skip over the product placement like they can commercials or leave the room when the commercial break is aired. Further, seeing contestants or stars using the product makes the product more credible to the viewer, especially when the viewer identifies with a contestant or has an affinity for the star in the first place.

How Product Placement Works

When a television program is being developed, often the content producer will look for ways to include products that are relevant to the program into scenes during each episode. It may be as subtle as a brand name showing on a computer that is being used or a product billboard that is recognizable in a scene. It may also be as obvious as the ever-present "Diet Coke" glasses with the Diet Coca-Cola labels facing the camera that the judges on *American Idol* drink from, to the Subway sandwiches that figure into the diets of *The Biggest Loser* contestants, to the weekly Ford music video on earlier *American Idol* shows where the contestants are seen enjoying and frolicking around a Ford automobile.[9] Regardless of the subtleness or lack of subtleness, product placement can be a significant source of continuing revenue for the content producer as the program unfolds.

While the aforementioned examples are from OTA network television, streaming services can also be successful sources of product placement. Products used in movies and original series on Netflix, Amazon Instant Video, Hulu/Hulu+, YouTube, and other streaming services are also examples of product placement. These days it is hard, if not impossible, to find a program on television – broadly defined – that does not have some form of product placement at some point during the program.

Product placement opportunities are generally set up through an agreement between the product maker and the content producer or the content deliverer in such a way that the product maker pays a set amount for the exclusive right to have its product displayed during the program. By exclusive right, the author means that a company would not agree to have both an Apple MacBook and a Dell Windows-based computer in the same program so that they would be competing against each other. Having two products competing directly against each other would negate the effect of the product placement of both products unless the product that has the contract is being shown to be far superior or preferred over the competitor.

Sometimes product placement is used as a barter opportunity. A television content producer may need products to offer to contestants or the public (such as game shows), or may need a product for the company, staff, or actors. If so, the content producer may choose to swap the placement of the product in the program for the product itself rather than to have to buy the product outright. No money changes hands and the content producer receives the product for use or distribution while the company producing the product gets the built-in advertising opportunity(ies) in a way that viewers will not be able to avoid by fast-forwarding or leaving the room as with commercials during breaks.

Second, product placement can combine barter with a promotional fee. In this scenario, the television content producer will swap product placement opportunities for the products themselves, but will also charge a fee for the promotional announcement. Often, these types of product placement are seen in programs with stable, and most likely, large followings, either in the total aggregate audiences or in the audiences most desired by the company considering the product placement opportunity. The promotional fee charged by the content provider will be less than the cost of purchasing advertising time in the program.

A third option is a sponsored deal – what, today, is called branded entertainment. In this option the product maker pays a fee to sponsor the telecast of a program. Depending on the agreement, the product maker will have its name associated with the program as well as have product placement opportunities located within the telecast of the program. Usually with this type of agreement, the product maker will also have a number of commercial spots during the telecast as well as the sponsorship in order to keep the product maker's name before the audience during the commercial breaks. Additionally, depending on the agreement, the product maker may also have significant input into the creative aspects of the program. This type of agreement was used extensively in 1950s television programs, and is resurfacing today, with Ford Motor Company creating its own reality program centered around its Ford Escape model vehicle. The program began running the last Sunday of March 2012 during prime time on NBC, and Ford tied the program in with its social media platforms.[10] While *Forbes* magazine reported that the program "garnered a ratings pittance" and was "no more compelling than a Ford 'music video' on *American Idol,*" the magazine also said that the program "takes Ford further down the path of branded video tied in with social-media platforms, where it has demonstrated industry-leading mastery."[11] *Forbes* points out that the program "is charged with showing off the new Ford model at every point that makes sense in the 'plot' and to viewers' tolerance for strong doses of product placement."[12] Crystal Worthem, brand content and alliances manager at Ford, said about the new reality series concept for the automaker, "It's exciting because, unlike other reality shows that shoot months and months in advance, we're actually shooting and airing in the same week. A crucial thing in the show is online interactivity, and the only way to get that is to shoot quickly."[13] The programs were taped weekly, but were done so in real time for that "live" feel. In addition, Ford did not actually buy the time slot from NBC, but instead decided to turn over all the ad revenue to the network in exchange for the opportunity to telecast the program. Ford's innovative strategy for the 21ˢᵗ century – actually very similar to the 1950s prime time game shows – was on display for six episodes Spring 2012.[14] In at least the new Ford television program, it seems as if 'what goes around, truly comes back around."

A fourth option for product placement is a straight payment deal. In this option, the product maker simply pays for the right to have the product placement opportunities during the program. No form of barter, where the content producer keeps the product in exchange for the placement occurs, so, usually, the product placement is a placement of the product within the scene of the program. Examples would be a billboard with the product's name and image prominently displayed within a scene. The product maker has paid for the opportunity to have that product placement in the program.

Why Product Placement Works

There are a number of arguments that you can make as to why product placement works, some going back to the academic research of the 1900s. Without turning this section into a thesis on theory, nevertheless, let me offer two theories very quickly as arguments for why product placement works. I'll try to make these as clear as possible.

In his 1922 book, *Public Opinion,* journalist and author Walter Lippman wrote that the public doesn't respond to events in the actual world in any direct way, but they live in a sort of "pseudo-environment" made up of "the pictures in our heads," and the media play important roles in shaping those pictures and designing the pseudo-environment.[15] Following along on Lippman's ideas in their study of the 1968 presidential election and other follow-along studies, two researchers, Dr. Maxwell McCombs and Dr. Donald Shaw – both of the University of North Carolina-Chapel Hill – described the media as "agenda-setters, that the media were not very good at telling the public what to think, but were extremely good at telling the public what to think about.[16] In other words, the media's real influence with the public is in setting the public's agenda for what is important to them in their daily lives. While the original studies focused on the news media, it's easy to see how agenda-setting can explain why product placement is so successful.

From an advertising point of view, if the media in general – and television, specifically for our purposes – are, in the words of McCombs and Shaw "stunningly successful" at setting the public's agenda of what is important to them, then product placement, by its sheer ability to keep a product before the viewer time and again in a program would potentially have a tremendous impact on the viewer's buying habits. Advertisers are extremely concerned with both brand awareness and brand recognition – the fact that the public is both aware of a brand and recognizes the brand. Commercials on television can be skipped by the viewer using TiVo or a DVR, or can be ignored by changing the channel or leaving the room during commercial breaks. Product placement negates the ability to do either. Whether overt or more subtle, product placement can drive both brand awareness and brand recognition, and, by sheer weight of continued exposure, can influence buying decisions. Product placement is excellent for helping to set 21st Century Television viewers' agendas for what they choose to purchase and will choose to purchase in the future.

Additionally, product placement can have farther-reaching effects than just the ability to set the agenda for the television viewers' buying decisions. In 1940,[17] and again in 1948,[18] Paul F. Lazarsfeld led teams of researchers in studying how people made decisions, first in a presidential election (1940) and then with regard to buying decisions on a number of everyday items (1948). From these studies came the notion of what the researchers called the "two-step flow" theory of communication. The two-step flow theory says that the mass media have only an indirect influence on the public at large, that the media influence opinion leaders of peer groups and those opinion leaders then influence their groups.

Peer groups are everywhere. Except for those people who are truly perpetual loners, anyone can think of groups of peers they are associated with. Peer groups can be from a job, a church, a sports team, a school or university – anywhere and any way groups of people get together to share things they have in common.

According to the two-step flow, every peer group has at least one opinion leader to whom the other members of the peer group look for making decisions – on what to wear, how to style their hair, what type of automobile to drive, where to go out to eat – decisions about all aspects of life. Applying the two-step flow to television can explain how television has a direct influence on the opinion leaders of the peer groups who then exert influence on the other members of their peer groups who make up the rest of the public.

Product placement, then, has a crucial role to play in the decision-making of the members of the peer groups that make up the public. With its ability to keep the product name and image before the television viewer, product placement is an excellent way to influence the opinion leaders who will then influence the other members of their peer groups. Further, because the other members of the peer groups will be watching many, if not most or all, of the same programs that their opinion leaders are watching, the other members of the peer groups will also be aware of and recognize the brands when the opinion leaders decide to use the brands and influence the other members to do the same. Product placement also has the ability to strongly influence the opinion leaders because, as mentioned with agenda-setting, product placement cannot be ignored or skipped like commercials can. The products are integrated into the program and are seen throughout the show.

Taken together, agenda-setting and the two-step flow make a powerful argument for the success of product placement as a powerful influencer of people's buying decisions. Agenda-setting suggests that product placement, by virtue of its continuous integration into television's programming, can be "stunningly successful"[19] at influencing the viewers to consider seriously making buying decisions based on the products they see. The two-step flow, then, says that the opinion leaders in the public will take those buying decisions to their peer groups which then extend the influence of the product placement to the public at large.

On a more practical matter, product placement makes brand awareness and recognition more accessible in a viewer's memory. When people are required to "reach into their memory for a product, just as when they reach for a word, those that come to mind quickly, the ones that are at the top of the mental agenda, have a distinct advantage over those that only emerge after extensive dredging."[20] When a product maker uses extensive product placement on a television program, by seeing that product on screen in a television program time and again, the viewer is that much more likely to notice, recognize, and select the product when (s)he is shopping, and is more likely to make it a point to look for that product because, through product placement, that product is most accessible in the viewer's memory.

As mentioned earlier in the chapter, reality shows are the television programs most heavily-packed with product placement. In 2011, researchers Emma Ashton and Julie Houston surveyed 400 viewers nationwide for their Reality TV Insights Survey. The two researchers found that 94% of the viewers' purchasing behavior had "been influenced by what they have seen on a reality show. 60% of viewers have purchased something after they saw it on a reality television show."[21] Notice that the researchers were specifically looking at product placement on television. The study suggests the power of product placement for both influence and, even better, buying decisions of television viewers. Ashton says, "It appears product placement is a win-win situation for both the shows, and the advertisers. Television networks need the sponsorship to ensure production values of the shows are high to attract viewers and therefore make a profit."[22] She goes on to say, "Advertisers are definitely getting a bang for their buck aligning with reality shows and there is probably much greater potential than is in play right now, if advertisers can think outside the box."[23] Product placement works.

Future of Product Placement

Product placement will continue to be an ever-growing part of 21st Century Television. The possibilities for additional revenues through the use of product placement are too numerous to ignore. Content providers will continue to use product placement as a source of revenue for their programs, and advertisers will find product placement an ever-more important way of reaching the television viewer because the viewer simply cannot avoid a product placement like (s)he can a commercial. Additionally, the ability to target viewers with specific advertising through the use of aggregated targeted microadvertising, programmatic media buying and dynamic ad insertion discussed in the previous section, combined with what the author calls *Ubiquitous Product Placement*, or UPP, gives the content provider an extremely effective one-two punch that will allow television revenues to grow ever larger and to flourish.

Product placement on 21st Century Television will be significantly different from product placement of today. Today's product placement is in its infancy compared to the future possibilities of the advertising strategy. Twenty-first Century Television will provide a continuous stream of product placement opportunities for advertisers to reach their intended audiences. Sometime in the next five to ten years, product placement will move from the Diet Coca-Cola glass on the desk of the *American Idol* judges to Ubiquitous Product Placement, a situation where everything in every frame of the television program will be a product placement. However, lest you imagine a television world that looks like a NASCAR driver and his automobile, the product placement of 21st Century Television will be much more subtle, and therefore much more effective, than the NASCAR driver with his/her sponsors plastered all over the car and driver.

Ubiquitous Product Placement on 21st Century Television will be designed in the following manner. Everything in the program will be an opportunity for a product placement. Each item in every frame will be linked to a website of the product where the viewer can go to find out more about the product and to make a purchase. Given the ability and the usefulness of the connected television set to allow multiple windows to be open on the television screen, and the Internet's ability to deliver multiple simultaneous operations such as allowing the viewer to watch television, chat with friends, look up background on his/her favorite actor, director, musician – all at the same time – delivering numerous simultaneous product placement web links to a viewer is not only a distinct possibility, but an almost guaranteed reality on 21ˢᵗ Century Television.

For instance, you, as the television viewer, are watching the current edition of *American Idol* (to stay with the example from earlier). One of the judges, Lionel Richie, has on a jacket that you as the viewer like very much. With a quick wave of your hand to the lapel of the jacket and a squeeze of your fingers (no longer a mouse, but you could use one if you are old-fashioned), a second window appears on the screen giving you information on the jacket – who is the maker, what colors you can get the jacket in, what the material is, etc. – and, upon further exploration of the website, the location of outlets – both online and brick-and-mortar – that sell the jacket, the cost, shipping information, as well as providing you an opportunity to order the jacket immediately.

Additionally, you notice that Katy Perry's bracelet would look good on your spouse's wrist. With the same wave of the hand and squeeze of the fingers or a spoken word you are taken to a website in a third screen giving you information on the bracelet – who the designer is, whether is it gold, silver, or some other metal, what stones (if any) are on the bracelet along with their cuts and weights – and what outlets sell the bracelet, along with costs, shipping, etc.

Thinking back to one of the most prolific product placement programs, you have gone back in time and are watching *The Biggest Loser.* As you watch, you notice the treadmill the contestants are using. Selecting the treadmill sends you to a second window where you find the same types of product and purchase information that was described in the previous example of *American Idol*. However, on this particular *The Biggest Loser* episode, the contestants are having makeover week, so every outfit, every hairstyle, every makeup used, becomes a clickable link. In addition, each contestant is taken to a special location where (s)he meets the people (spouses, children, family members) the program coordinators have chosen for them to see. That special location will also be a clickable link, which, if you select it, will open a window to provide you with information about the location and any other pertinent facts you as the viewer may need.

Several weeks later, the contestants participate in *The Biggest Loser* marathon, where they show off their new-found abilities to run, walk, limp, or whatever they have to do to finish the 26.2 mile race. In 21st Century Television, every location the contestants pass, everything in each scene that would be of interest to a viewer will be a clickable link to take her/him to an additional window(s) for more information and potential purchase.

Summary

Much like addressable advertising on 21st Century Television, Ubiquitous Product Placement will be another option for advertisers to reach their preferred audiences. Initially, the ability to target subsets of the audience in the way that ATMA is able to, likely will not be available. However, as the sophistication of the content providers and the advertisers increases, the likelihood that some form of ATMA could be combined with UPP to place a Coca-Cola bottle in the streams of a subset that prefers Coke and a Pepsi-Cola bottle in the streams of a different subset that prefers Pepsi is almost assured. The opportunity to sell different product placements to different advertisers in the same program will further increase the revenues for the content providers.

Further, as audiences become ever more comfortable and familiar with Ubiquitous Product Placement in all their favorite programs, the expectation is that they will use the ability to immediately find out more about products and services that interest them. The desire by the audiences for instant gratification will lead to more and more impulse buying and increased revenues for the advertisers. In addition, as UPP increases in number and sophistication, content providers will once again, like with addressable advertising, be able to increase the price of UPP insertions by charging a premium for each insertion, because the advertiser will be assured that its product will reach the desired target audience.

When everything in every shot becomes a clickable link, Ubiquitous Product Placement becomes a continuous gold mine for everyone involved in the 21ˢᵗ Century Television program – the content producer, the content provider, the product maker, and the advertiser. The content distributor will benefit from 21ˢᵗ Century Television product placement by tying in paid messages to the product placements in the program being distributed. Even the viewer benefits by having instant access to anything (s)he sees on the television screen that draws his/her attention or interests her/him. In 21ˢᵗ Century Television, product placement is not just a "win-win" situation; it's a win for everyone.

CHAPTER 9

PROMOTION

One of the most important aspects of 21st Century Television is the need for continuous promotion by the content providers. With the demise of scheduling brought about by the growth of video-on-demand, and the requirement of 24-hour-a-day television everywhere and on all platforms, making sure that audiences can find the programs they want to see will be critical to the success of both the legacy and the new content providers.

As has been discussed in previous chapters, all content providers will need to have available huge libraries of programming for their audiences. The networks especially, will be expected to make sure that more than sixty years of programming is available to anyone and everyone anytime, anywhere, and optimized for every platform – including VR and windshield displays for autonomous automobiles. How audiences will know about, and how they will find these programs when they wish to see them will be a daunting task for all content providers.

The traditional cable channels, with their niche programming structure, will still have backlogs of programs that their specialized audiences will want to view whenever and wherever they wish to view them. No longer will programs show up one week on, say,

The History Channel, and several weeks or months later, appear on The American Heroes Channel (formerly The Military Channel) or The Discovery Channel. Instead, The History Channel will have its own library of programming that will be unique to that content provider and will be viewed only on that provider. The same will exist for the other traditional cable channels.

Promotion, then, will need to take a variety of different forms. First, a content provider will have to devise alternative ways of letting audiences know what programs are available from that provider. Without schedules, listings in *TV Guide* or *The TV Guide Channel* will be useless, because at each half-hour all programs that a content provider has in its library would have to be listed, and then the next half-hour the program guide would have to do the same thing. It would be a ridiculous waste of time and energy for the content providers, and a useless piece of information for viewers.

Instead of using the traditional ways of letting the audience know what programs are available for viewing, in the 21st Century Television universe, **filters** will become the choice of the content providers. Filters will make it possible for viewers to find what programs are offered by each content provider. The filters will work in the following manner:

Because the content provider's website will be its most important link to the audience, the website will have an area listing all the programs offered by that provider. Those programs will be cross-referenced in a variety of ways – alphabetically by program name, by character, by genre, by lead actor, by lead actress, by setting or locale, by time of day the program traditionally aired, or by any number of other ways of cross-referencing. For instance, daytime soap operas could be listed as "Early Afternoon" or simply "Afternoon." The list of potential ways to categorize the programs is extensive. The more ways a content provider can cross-reference its programs for the audience, the better. The viewer would need only to know any aspect of the program to be able to find it with just a few clicks of the mouse, a wave of the hand, or a verbal command. Further, because some categories would be exceedingly large (e.g. "Drama" would be a huge category), content providers would combine category filters with other filters (such as the ones mentioned above) to take the viewer swiftly and easily to the program (s)he wants to watch.

The 21ˢᵗ Century Television viewer will be able to easily, quickly, and smoothly navigate to his/her desired program without a second thought. First, the viewer would log in to the content provider's website. The log-in procedure allows the content provider to know the viewer and to build a continuing profile that allows the content provider to deliver the preferred programs and most effective advertising to the viewer. Then, for example, if a viewer wanted to watch, say, the third episode of the second year of the program *The Big Bang Theory,* a CBS Network situation comedy, the viewer could simply go to the CBS Network website, click on "Library," then "Sitcom," then *"The Big Bang Theory,"* then select the year, then the episode. Or, after clicking on "Library," the viewer could simply click on *"The Big Band Theory,"* then select the year, then the episode. Another way would be simply to click on "Library," then "Sheldon Cooper" which would then bring up *"The Big Bang Theory,"* then select the year, then the episode. The viewer could also select "Library," then "Jim Parsons," then *"The Big Bang Theory,"* then select the year, then the episode; or the viewer could select "Library," then "Kaley Cuoco," then *"The Big Bang Theory,"* then select the year, then the episode. In the quickest way, the viewer could just search for *"The Big Bang Theory,* Year Two, Episode Three." You get the picture.

Once all 21ˢᵗ Century Television sets are fully motion controlled, the mouse clicks referred to above would be replaced by a wave of the hand for each click. Further, when the television sets are fully voice activated, the only commands to watch the *The Big*

Bang Theory episode in the example would be, first, "Library," then "*The Big Bang Theory*, Year Two, Episode Three." Using voice commands or motion control for selection will be much easier and will eliminate the need for the extra hardware.

While inputting the various links would be a very large task at the outset, once the cross-reference scheme has been devised and implemented, it would only have to be updated as new programs are added. Furthermore, because all programs have value, they will also have advertising in every showing, connecting advertisers with the audiences. After the initial setup, all of the procedures will occur seamlessly through the cloud service controlled by the content provider.

Additionally, because for 21st Century Television interactivity is the key, while the viewer is watching his/her program, there will be recommendations for other programs that the viewer might enjoy, based upon the viewer's program choice or profile. Those recommendations will be located in a different portion of the television screen and will have links to each of the suggested programs. Further, as the viewer watches a particular episode of a program, (s)he will be asked whether or not (s)he would like to include that program or even that episode in a "Favorites" category, which will make it easier for the viewer to return to in the future. On the home webpage, the viewer can then access his/her "Favorites" category and go straight to those programs (s)he has included. Early simple forms of this type of recommendation are occurring on television today as the networks and cable channels will often highlight coming programs by the use of small static and animated forms in the different corners (usually lower left and lower right) of the screen. These recommendations will highlight the next program to appear on the network or channel, a program airing later in the day or a later day, or – in the case of a family of channels – what will air on another of the family's channels. Twenty-first Century Television's recommendation system will be dramatically more sophisticated and higher quality than the early examples of today.

A second way the viewer can find the program (s)he wants, will be to work through a series of filters on the major search engines such as Google, Yahoo, MSN, etc., on new search engines devoted exclusively or primarily to television programming, or on social media sites. The search engines will be used primarily when a viewer knows the name of a program or some information about the program, but doesn't know the content provider that has the program in its inventory. When using one of the search engines, the viewer will insert any portion of the name of the program, lead character, lead actor/actress, etc., and with only a click or two, wave of the hand, or voice command, be at the content provider's website in its library, at or near the program the viewer wants to see. The process would occur in a manner similar to the *The Big Bang Theory* example given previously, but with the search engine as the beginning point. The more information regarding the program or episode that's provided at the beginning, the fewer clicks, hand-waves, or voice commands would be needed.

Additionally, the specialized search engines designed for television specifically will make it possible for the viewer simply to input a particular aspect of the program into the search engine. The search engine would then take the viewer directly to the listing or near the listing (depending on what is inputted), with little to no intermediary steps to have to navigate.

But what about new programs? Production houses will be running at full capacity to keep up with the demand for new programs from the ever-proliferating universe of 21ˢᵗ Century Television content providers. The multitude of new programs must be promoted and promoted heavily if they are going to succeed with the viewer and with the advertiser. The most obvious place for promoting new programs will be on the content provider's website. Because scheduling no longer exists, or the need for "seasons," new programs can be introduced continuously on the website. Further, like the suggestions of alternative programs on the screen while the viewer watches a particular program, on a portion of the screen will be blurbs about the new programs that are available for the viewer to watch, with clickable links to "jump" the viewer straight to a new program's episode. There could even be a section on the television screen where the new program titles (as clickable links) for the particular content provider will scroll so the viewer can see the listings and choose to "jump" to a new program right away. Alternatively, there could be a section of the screen where, instead of a list of new programs by the content provider scrolling, there could be a list of programs of the same genre or with the lead actor/actress or with the same or similar setting, etc., regardless of content provider, that would scroll for the viewer to consider and possibly select. The choice of how the scrolling list is compiled could even be left up to the viewer's preference.

With 21st Century Television, there is no problem with the viewer jumping to the new program episode from the one (s)he is watching. If every program is available all the time, the viewer can jump to the new program, watch it in full or in part, then jump back to the program (s)he was originally watching. Viewers will also be able to enjoy the same opportunities on their tablets, computers, and smartphones, although most likely in a more basic version of the website, depending on the device. While, at this time, it may seem as though there will be numerous distractions on the screen as a person watches a program, 21st Century Television viewers will consider this type of viewing normal, especially the younger age groups for whom multitasking is just a way of life. In fact, as discussed in the chapter on The Viewers, the Deloitte study shows that Millennials and Gen Z's already multitask using as many as three or four different devices simultaneously.[1]

A second way of promoting new programs is to the viewer directly as (s)he is watching television. This second way may actually be better than the "shotgun" approach of placing all new shows on every page. Here is where knowing the viewer is extremely valuable. Once the viewer has logged in to his/her account on the website, all of his/her preferences, likes and dislikes, demographic and life-style information, as well as television habits and previous choices become available to the content provider to help determine which of the new programs being introduced the viewer is most likely going to want to watch. As soon as the logged-in viewer makes his/her first selection for the program (s)he wishes to watch at that particular time, blurbs for new programs of the same or similar type will begin appearing on the screen to entice the viewer to tune in and try the new programs out. A simple click, wave, or verbal command will take the viewer directly to that new program, or save it to a "New For Viewing" category if the viewer wishes to watch it at a later time.

It is important to remember that the viewer is choosing what (s)he wishes to watch, **when** (s)he wishes to watch; therefore, because all programming is on-demand, it doesn't matter whether the viewer jumps to the new program or not. If (s)he does jump to the new program, after the viewer finishes watching the new program, (s)he can simply return to the program the viewer was originally watching. If the viewer saves the program for later viewing, then (s)he can watch the new program at his/her leisure. Regardless, the program, with its included commercials, gets watched.

Further, if two or more people are watching television together, and both have logged in, then blurbs for new programs designed for each viewer will be displayed on the screen, and those programs can be saved to each viewer's "New For Viewing" folder. The viewers can finish the program they are watching together before going to one of the saved programs, or they can jump to a new program that one of the viewers wishes to watch. It will even be possible – especially in dealing with children and the desire of each child to watch his/her own program **now** – with a television screen large enough, to split the screen into multiple smaller ones at the same time, one smaller screen for each child. Such a showing of multiple programs would require a separate headset for each child, providing the correct audio for each child's program to the child.

The use of split screen or multiple split screens would also be available for adults, when the need arose. For instance, as described earlier, two or more viewers are watching a program together and they are all logged in. Because they are all watching the same program together, each time a blurb for a new program appears on the screen, any or all of the viewers will have the opportunity to save the program into the "New For Viewing" folder, regardless of who the program blurb was designed for. Multiple savings of the blurb for the new program makes it possible for the content provider to reach more viewers in addition to the intended viewer, allowing the content provider to increase the potential audience for the new programs. Further, if one or more of the viewers (but not **all**) decide to watch the new program immediately, by selecting the new program, additional windows would open on the television screen for each viewer who wants to watch the new program. Headphones, then, would be needed for each viewer to receive the audio from either the new or the original program.

New program offerings can be shown on the various pages of the major search engines. In the future, there will be more search engines than the current ones or the major search engines will offer a variety of different, specialized search engines, depending on the category and the need of the category to have its own search engine. Search engines devoted specifically to television programming will become the *TV Guide* of the future, providing the 21st Century Television viewer with the ability to filter the programs in the ways described previously. Such search engines or competing search engines devoted to television programming would eliminate the need for Google or Yahoo or MSN or other general search engines to include the extra link(s) necessary to go from the general search engine to the program. Links only to the specialized search engine would be necessary, and those links would only be necessary for a short transitional period until viewers became accustomed to the television search engines. The television search engines would function much as a regular search engine, but with the convenience of providing links only to television programs. Ads, promotional materials, news blurbs (if desired), and other informational content could be included on the search engine pages, and logged in viewers using the search engine could be prompted to their favorite types of programs up front with links to those programs readily available to the viewer. Additionally, international search engines – either general or specific to television – would provide opportunities for content providers to reach audiences around the globe. Having their libraries of programs listed on international search engines would make those programs even more readily available to new viewers and new audiences in general. Reaching global audiences, then, opens numerous opportunities for advertisers to reach potentially massive numbers of new consumers for their products in addition to the huge numbers of new viewers for the content provider's programs.

In addition to the use of search engines and filters to promote both new programs and programs already in the library of the content provider, social media will become increasingly important for promoting a content provider's programming. Facebook, Linkedin, Twitter, SnapChat, Instagram, Pinterest, and other social media sites will offer the television content providers numerous opportunities to promote their programs directly to their "friends" and "fans" via pages, likes, tweets, etc. As users of the social media sites provide more and more personal information about themselves, programmers will be able to target their "friends" and potential friends with information and content designed to drive viewers to programs and especially new programs. When a new program or a new season of a current program is about to "hit the air" (is it even relevant to use that term any longer?), Twitter, Periscope, Facebook, and all the other social media machines of the production house's and the distribution company's promotion departments will begin firing tweets and other short messages and videos to those with connections, letting them read short, promotional blurbs and see short videos from the producer, director, the stars, and the distributor about the wonderful new program or season that will be premiering. Additionally, the production house and the distribution company will post on their social media sites similar promotional blurbs for their friends and fans to comment on and to "like." Those friends and fans can then pass along to their friends, information and comments about the new program in a continuous web of promotion. Further, on social media sites, the production house and the distribution company will post short promotional videos of the program designed to whet the appetites of their friends and fans who will ultimately become their viewers, much as movie trailers today can be posted on such social media sites.

Once the program has been purchased by a content provider, the provider's promotions department begins its work. Much as described earlier about the production house and the distribution company, the content provider will have the same variety of social media sites to promote its programs to the viewers and potential viewers. The provider will begin with a series of tweets on Twitter, status updates on Facebook and LinkedIn, short, streamed teasers on Periscope and Instagram, and other such beginning promotional announcements on additional social network sites. Messages to "friends" and "fans" will get the pattern going as those supporters or just viewers begin to tweet and "statusfy" about the new program or programs being offered.

Viewers will update their status on Facebook or other social media sites, stating their feelings (hopefully, positive feelings) and re-posting the short video promos. Others will see the positive reaction and interest in the new program by their friends and the friends-of-friends and will begin talking about and re-posting the videos and statuses about the program. Each Facebook status update, video post, tweet, or other addition to the social media sites by a viewer or a friend of a viewer has the opportunity to reach more and more potential viewers of the program. These promotional tools will begin to generate "buzz" (the word-of-social-media, an alternative to, and, increasingly, replacing, word-of-mouth) that will yield significant numbers of viewers for every program in the content provider's inventory, not just the new program that got the discussion started. The same thing will occur simultaneously on all the other social media sites, expanding exponentially the opportunities for the content provider to drive viewers to the program.

Further, because the program would be telecast over the content provider's website, the potential to draw viewers to the new program in international locations would be virtually certain, increasing the total number of viewers of the program. At the same time, the promotion work will encourage viewers in international locations to explore further down the "Long Tail" of the content provider's inventory of programs. These programs will often find favor with international viewers in different locations, especially if they have likely not seen the programs due to the lack of international distribution efforts, before the development of 21ˢᵗ Century Television. The "word-of-social-media" promotional efforts by the production house, the distribution company, and the content provider will open numerous new opportunities for classic programs to find new life internationally, with new viewers and new opportunities for advertisers to reach existing markets, or to develop new markets around the world. Those opportunities, then, become new sources of revenues that will enhance, and, potentially, could eclipse the revenues generated domestically.[2]

Additionally, the social media pages of the content producer, content distributor, and content provider will be involved in even more engagement forms of promotion. Friends, fans, etc., could have the opportunity to watch program clips and rate the program even before its first episode aired. In addition to promoting the program, such responses could provide feedback on the potential popularity of the program once it began to be aired, although, once a program is in the inventory of a content provider, it will cost virtually nothing to continue to provide opportunities for the program to be seen. Even a program that does not begin with a strong following, could, over time, develop a large or at least loyal following. In other words, the days of allowing a good program to grow into its audience (if no longer its time slot) would return. Further, by using social media as one of the major promotional platforms, the content providers could develop contests for their friends, fans, and potential friends and fans to participate in, thus building interest in the program and in the content provider while providing additional ways for viewers to become engaged with the provider.

Additionally, the social media pages would be used as a way to introduce friends, fans, etc., to the actors and actresses that would be starring in the new program. The stars will tweet, update status, etc., on all the various social media sites, on their backgrounds (especially if they are not well known), become "friends" with the viewers and potential viewers of the program, and direct the viewers and potential viewers to the actors' and actresses' personal websites. In that way, viewers and potential viewers could develop a relationship with, and an affinity for, the actors and actresses. By building viewer interest in the actors and actresses, the content provider would be enhancing its own closeness with the viewer. All this "closeness" would encourage the viewers and, especially, the potential viewers, to "check out" the program when it aired, and would give the viewer even more reason to return again and again to watch the content provider's programs being offered to her/ him with his/her profile in mind.

By providing profile information on their various actresses and actors to their social media fans, the content provider also will provide a means for their fans to access those actors/actresses own social media fan pages. There the fans could learn even more about the actor/actress, gaining background information about the star, his/her likes and dislikes, what other programs (s)he has been a part of (thus building interest in other programs the content provider might have in its library of programs), what the star thinks about the current program (s)he is in – anything that might heighten the interest of the fan in the star and thus in the program.

Social media can be one of the most powerful forces available to the content provider in reaching the viewer in an in-depth way, and, in developing such a rich relationship with the viewer and the potential viewer, the provider can make available the exact programming that will be effective in capturing and holding the attention of the viewer. While much of what is described here is already being done at least to some degree, promotional work using social media sites still has a long way to go compared to the possibilities of what will be able to be accomplished in the future.

Last, but certainly not least, there is YouTube and other user-generated video sites, along with the decisions by Twitter and Facebook to include videos on their sites as well. Already, content providers are experimenting with putting content from their programming onto YouTube and other such sites in an effort to drive viewers to their programs. While much of the current television programming content placed on these sites is, in many ways, the first steps in the right direction, as program delivery changes, such sites will be primary tools for promotion that the content providers will use to get the word out about the programming they have available.

One of the best uses of YouTube and other user-generated sites will be to promote new programs as they come online. In addition to the other promotional opportunities mentioned above, user-generated sites will serve as excellent tools for promotional campaigns for the content providers' new programs. Further, because YouTube has branded channels, those channels can be used by the content providers and content producers to "test out" new programs, new talent, new plots, etc.[3] Additionally, YouTube itself has begun to deliver its own original content through its channels, so that becomes an additional channel for previewing programs before they move to the more mainstream content providers.[4]

Weeks before the new program is to debut, short promotional videos designed to entice viewers to the new program will air on the user-generated sites, and audiences will be directed to the video samples through the content provider's website and its social media pages. These videos will not only be designed to excite the viewer to watch the new program, but the user-generated sites will offer the content provider the opportunity to run videos that will have added features. These features can include interviews with the stars of the program, the director, the producer, and possibly testimonials from viewers who have had a chance to participate in viewings and screenings done prior to the promotional campaign.

In conjunction with the user-generated site campaign of video material, the content provider will begin a "kick-off" promotional campaign for the new program on its website. The campaign will tout the new program, its stars, its debut date and time, **and** the opportunity for viewers to get a first-hand look at the new program, along with behind-the-scenes information of the new program by going to the user-generated sites where the campaign is occurring. The content provider then begins a social media campaign, getting the news of the new program to its friends and contacts, urging them to check out the user-generated sites for the new program and extra material they can find there, and passing along their thoughts on the new program to their friends. Additionally, given the extensive knowledge that the content provider will have about its viewers, the promotional campaign can be targeted to those viewers who will most likely be the program's primary and secondary audiences.

As the date of the new program's debut grows close, more enticing videos are placed on the user-generated video sites, either in addition to, or replacing the earlier videos. The content provider's website begins touting the new program and promotional spots are placed on the search engines, especially for programs expected to be blockbuster-type hits (but not necessarily for extreme niche programs unless the spots are on niche search engines dedicated to that type of program). The program's stars begin serious and extensive use of social media sites and tweeting with their friends, encouraging them to watch the program. What is produced is a highly organized, extensive, multifaceted promotional campaign that will be extremely effective in driving preferred audiences to the new program, in a promotional campaign that will be not only cost-effective, but virtually cost-free. The cost savings compared to the extremely expensive promotional campaigns of today's television in promoting new programs will be enormous, while the results will be significantly more positive.

In addition, advertisers can assist in promoting the content provider's new program. Advertisers expecting to run commercials in the new program could begin adding promos for the new program to the home page of their websites, with encouraging comments about the program and links to the different video sites where promotional videos are occurring and links to the content provider's website as well. Given the potential for huge revenues from 21ˢᵗ Century Television, it will be in the advertisers' best interest to work closely with the content providers in promoting the programs they plan to place commercials on.

The promotion of new content will not be the only use for such sites. Content providers will find that they can promote their entire inventories of programming over these sites. By promoting whole libraries on such sites, content providers will be able to reach new viewers who have never even known of such classic shows as *I Love Lucy, The Jeffersons, The Ed Sullivan Show,* and *Rowan and Martin's Laugh-In.* By promoting those older (and even ancient in the minds of the youngest viewers!) programs to these new audiences, content providers will have the potential to pick up additional viewers for whom those programs would be a fascinating new treat.

Likewise, the promotion of programming over user-generated sites would also reach those older viewers for whom such classic shows as those mentioned above would hold fond memories. Often, in the bombardment of ever-new and ever-changing programming, classic shows are forgotten until something stirs an idea in the mind. YouTube and other user-generated sites would serve as the catalyst needed to stir those memories and drive viewers to those programs.

For instance, CBS finds that the audience for *I Love Lucy* is beginning to level off or shrink from its previous viewer numbers (however large or small those numbers may be over a viewing period). Using YouTube and other user-generated sites, CBS begins a campaign to draw viewers to the program through a promotional campaign aimed at both new viewers and classic viewers of the program.

In its campaign, CBS places on its YouTube channel certain classic episodes of the program, along with classic comments and interviews (should they exist) from the stars, producers, directors, etc. Then, on the CBS website and through social media sites, CBS begins a campaign urging viewers to "check out" the "classic hit series from the early days of television, *I Love Lucy.*" Viewers going to the YouTube site could then watch the episodes with limited, or no commercial interruptions (remember, this is a promotional campaign), then comment on what they have seen. CBS could even make it possible for the viewer to interact with his/her friends immediately after or even while the episode is running to encourage those friends to watch and chat along with the viewer.

The use of YouTube and other such sites as promotional devices to drive new audiences to classic programs, as well as driving familiar audiences to those same classic programs yields an additional benefit – the ability to deliver advertising not only to a whole new generation of viewers but also to familiar audiences, both for virtually no cost, and thereby developing additional sources of revenue. While the audiences may be small overall compared to new programming and current programming at any one time, classic programs and those that will become classic programs as time progresses will still have their audiences and will still be able to deliver revenues. With virtually no costs, these programs could provide comparably substantial revenues.

YouTube offers additional opportunities for content providers to promote and even air special programs – possibly on a trial basis, but not necessarily so – on a YouTube branded channel belonging to the content provider. Alternatively, given the percentage of programs the networks, especially, own and telecast today, the content providers of 21st Century Television could choose to air original programs they have produced and/or own on YouTube as original content. The opportunities for the content providers to reach millions of viewers in a short period of time on the best-known of the user-generated channels would be excellent for testing new program concepts and ideas, introduce and test new genres (or reinvigorate otherwise "dead" genres), and give new writers, producers, directors, and actors/actresses opportunities to develop loyal audiences before moving to the content provider's website and inventory of programs.

Summary

Promotion will become even more important to 21st Century Television when there will be no scheduling and everything will be on-demand, anytime, and anywhere. Websites; search engines, both current ones such as Google and Yahoo and future ones devoted strictly to television programming; social media sites; and user-generated sites will be the sources of information for 21st Century Television audiences. Those audiences will find not only the newest programming being offered on all of the different content providers' channels (websites), but also the current and even classic programs that are available to them.

Promotional strategies and techniques will make viewers aware of what's available and what's soon to be available. It will inform and entertain the 21st Century Television viewer, while it leads her/him to the programming (and, by extension, the advertising and product placement) that is right for each viewer. In short, promotion will be critical to the success of every 21st Century Television content provider and every viewer – in short, everyone.

CHAPTER 10
MONETIZING THE GLOBAL MARKET

S o far, this book has taken a look at the major players that are a part of 21st Century Television, the differences in the television viewing age groups as television moves into the 21st century, and the revenue streams and promotional developments available for participants in 21st Century Television. The final piece of the puzzle is to look at the idea of 21st Century Television as global television. While other chapters have mentioned the global nature of 21st Century Television, this chapter explores the possibilities. While addressable advertising and ubiquitous product placement will produce revenues that can satisfy the money brought in by traditional advertising, it is the global nature of 21st Century Television combined with addressable advertising, ubiquitous product placement, and continuous promotion that will usher in a new "diamond age of television," the likes of which will surpass even the wildest dreams of today's television industry. It is the development of a global television industry that will become the future of 21st Century Television.

We live in a global society. Since the end of the Second World War, the world has continually seen a move away from isolationism and a move toward the globalization of the entire planet. No longer can a country – regardless of the size or the abundance of its natural resources – successfully choose an isolationist stance. Those countries that have chosen isolationism have seen themselves relegated to a role of irrelevancy and often pariah-nation status. The question these days is not whether to go global, but how to make it happen.

Interestingly, United States television has traditionally been a nation-specific set of industries. U.S. television, whether over-the-air or cable, has, for the most part, been limited to the fifty states, plus the incidental spillover that occurs along the U.S. borders with Canada and with Mexico. For subscribers to DirecTV and Dish Network, their international reach has been ancillary to their U.S. subscribership. Their international reach is limited to the Western Hemisphere, and has, for the most part, been aimed at the immigrant population in the U.S. who wish to still connect with their home countries in their native languages. While television programming from the U.S. travels the globe very well, showing up on local stations in many, if not most countries of the world, the networks themselves focused only on delivering their programming to the U.S. television audience. That, however, is changing.

Through the use of satellite delivery, the networks – especially Fox and NBC – along with a number of cable channels such as CNN, Discovery, and ESPN – to name only a few – have been moving to open the rest of the world to their programs, either in English or in the local language in some form. CNN, of course, in many ways started the internationalization of the U.S. television industry when it began its international telecasts with CNN International in 1984, an English-language news channel provided to upscale hotels, and aimed primarily at American and other English-speaking-or-understanding businesspersons and tourists.[1]

Today, viewers around the world can see a number of different programs on specialized American television channels if they have a home satellite system or cable. These specialized channels generally are not available in the U.S.; they are designed for the international audience, although they often telecast programs that are also shown to the U.S. audience. Depending on how the television channel chooses to distribute the programming, audiences might hear the English language version, an overdubbed version usually in a dominant language of the region or country, a local language version, or an English language version with subtitling. Advertising often is country-specific or regional in scope, and can be considered in the same category as local advertising.

Please allow me a few personal examples. I have had the opportunity throughout my career as an author and a university faculty member to travel extensively. As such, I have had the chance to watch a variety of television channels in countries in Europe, Central and South America, Asia, and Oceania. I have been able to watch one of the CNN channels regardless of the country where I was at the time. I have had the opportunity to watch the NBC SuperChannel, CNBC, Fox Sports, live college football bowl games, tape-delayed NBA basketball games, the occasional MLB game of the day, and even on rare occasions, NFL games. I have had the chance to see Discovery Channel programs around the globe. Further, I have had the opportunity to watch my favorite scripted television programs on a delayed basis in virtually every country I have visited.

To turn the situation around, I also can watch BBC America on my DirecTV system, BBC programming on PBS stations, and, at one time, news programming from Canada. Further, depending on where I am in the U.S., I have had the opportunity to see programming from Japan, the Middle East, and Mexico on the television set in my hotel room.

One more turn of the situation – in my travels around the world I have had the opportunity to watch television channels from countries other than the one I was visiting or doing business. Satellite television has made it possible for pan-European and pan-Asian delivery of channels, both country-specific and those that are truly pan-continental.[2]

The point is that television – both the programs as well as individual channels – is becoming more and more global. Television is no longer a local, regional, or even national entity. Television has transcended national boundaries, and has made itself into a global commodity, taking with it programming genres, sports leagues, and niche channels. Twenty-first Century Television takes television into the future, a future where all television is global television. The possibilities are virtually limitless.

21st Century Global Television: TV Everywhere

Twenty-first Century Television is truly global television. Twenty-first Century Television, as described in the previous chapters, makes it possible for every content provider potentially to reach every Internet user on the planet directly with programming, advertising, product placement, and promotion. That's a potential audience of almost 4.4 billion people around the globe as of March 2019. That's 56.8% of the total world population, and a growth of 1114% since the turn of the century.[3] By geographic regions, Asia has slightly less than 2.2 billion Internet users (a 51.8% penetration rate – the first time ever that Asia has gone over 50% - and a growth rate since 2000 of 1822%); Europe has almost 720 million (86.8% penetration and 584% growth); North America has more than 327 million (an 89.4% penetration and 203% growth); for Latin America and the Caribbean basin it's more than 444 million (67.5% penetration and 2360% growth); for Africa it's almost 493 million (37.3% penetration and an amazing 10,815% growth rate); the Middle East has more than 173 million (67.2% penetration and 5183% growth); and Oceania/Australia has more than 28 million

Internet users (68.4% penetration and 276% growth).[4] Each one of those almost 4.4 billion Internet users worldwide becomes a potential viewer of 21ˢᵗ Century Television and a consumer to be reached.

The National Association of Broadcasters is the largest organization of its kind in the world. Every year, more than 100,000 participants converge on Las Vegas, Nevada, to be a part of this convention. One-fourth or more of those participants come to the convention from outside the U.S., and the numbers are growing. Yet, time and again, in his convention-opening address, NAB President and CEO Gordon Smith at best mentions the global participants only in passing if at all. Yet the participants from around the world still enjoy the several days of the convention because they realize that the future of television – 21ˢᵗ Century Television – is a global television universe, and they have a part to play in this universe. Indeed, many of the countries around the world play major roles in the development of 21ˢᵗ Century Television, even taking the lead and showing the U.S. the possibilities. For instance, South Korea is leading the world in the development and implementation of ATSC 3.0, rolling it out to the entire country and delivering its television entirely through ATSC 3.0 by 2027.[5] ATSC 3.0 in the U.S. is still in the development stage and it remains to be seen if it will even become a reality and a serious competitor to the other IPTV-based distributors.

This "lip service only" is not new. In his opening address to the NAB Convention in 2012, Smith's stated, regarding ubiquity in broadcasting, that in the previous days of television, ubiquity described a desire to have a television set in the living room of every home, and that today (2012), ubiquity describes the idea of a television set in almost every room of the house.[6] Further, to quote NAB President Smith directly from the Jessell article, "But ubiquity tomorrow must mean broadcasting's availability to all people at all times *in all places* and on all devices."[7] (emphasis added) Smith further said that broadcasters "need to be on tablets, laptops and game consoles and on mobile devices not yet developed."[8]

While Smith was on the right track, he never included in his view the idea of making each signal a global signal. In his subsequent addresses to the NAB, Smith has continued to focus on the developments in the U.S., and the pride the U.S. broadcasters should have in what they have accomplished and are accomplishing. While it is understandable, given the make-up of the convention, by not including the global nature of the future of television in the coming years, this lack makes Smith's views short-sighted and, for the most part, out of date.

The future of television is not country-specific, nor is it country-specific with bleed-over on the borders. Rather, 21st Century Television is a global enterprise with ownership of content providers crossing national boundaries to become global enterprises. In some ways, television is already a global industry. Satellites beam television signals from anywhere in the world to anywhere in the world. Syndicators sell programs across the globe. Comcast, Charter/Spectrum, and others, through the development of TV Everywhere, are making their programming available to their subscribers any time, on any platform, and – literally – everywhere a compatible cellular or a Wi-Fi signal is available, anywhere in the world.

In the future, the television audience will demand even more – and 21st Century Television will provide. With the move by television to the Internet through the use of IPTV, content providers will no longer be limited by channels, local affiliates, or even national borders. Netflix is today demonstrating what can be accomplished as it is currently in more than 190 different countries around the world, or virtually every country in the world, and it is currently negotiating with China for access there.[9] The 21st Century Television universe will provide the viewer not only the possibility of television anytime, anywhere, and on any platform, but also any program regardless of its point of origination in the world.

With the development of global television as the standard fare for the 21ˢᵗ Century Television viewer comes new opportunities for revenues for the content providers. Combining Aggregated Targeted Microadvertising, Programmatic Media Buying, and Dynamic Ad Insertion conducted worldwide, with expanded, ubiquitous product placement, gives the content provider the opportunity to develop numerous revenue streams by delivering different advertisements during the same commercial insert time and different product placements to viewers around the world. By having the entire inventory of a content provider on its home website or a separate, specialized website, 21ˢᵗ Century Television content providers will be able to reach their worldwide viewers with advertisements and product placements no matter who they are or where they are in the world. The opportunity to provide programming seamlessly to viewers around the world opens television to an entirely new universe of opportunities for ever-increasing worldwide revenues that makes the future exceedingly bright.

Twenty-first Century Television viewers want their television anywhere, anytime, and on any platform – that point has been made again and again. A possible subtitle of this chapter – "TV Everywhere Fulfilled" – makes the point that everywhere means exactly that – the viewer is not limited by geography or national boundaries or even oceans. It doesn't mean that viewers can watch their favorite programs anywhere, except for the broadcasters whose online signals must stay in their own markets. For the networks and their local affiliates, that's a suicide plan.

Ultimately, though, the networks won't sign on to a plan that guarantees their ultimate demise. As discussed in Chapter 2, the networks will not go down with their affiliates, choosing instead to compete with all other content providers by delivering their programming directly to their viewers. This decision, which will occur almost certainly by 2025 (and possibly as early as sometime in 2020), will be driven by the networks' continuously expanding use of cloud technology for storage of their program inventories, along with the ability to deliver their programming through IPTV, be it wired, Wi-Fi, cellular, ATSC 3.0, or some other form yet to be developed. The decision to drop their affiliates also removes the networks from the burden of non-overlapping geographical markets as well as national borders with their spillover concerns, and allows them to reach their audiences anywhere, including any location outside the United States.

The networks are already making their first moves in this direction, and, in a sense, are already global broadcasters. Currently, they make it possible for viewers everywhere to watch archived episodes of programs on their websites and by providing apps for smartphones and tablet computers that allow viewers to watch streamed programs directly on their devices.[10] What the websites and the apps currently lack is the ability to deliver programs when they are originally scheduled, but that is a political decision and not a technological one. For the networks, taking the final step of airing their programming online at the same time as their affiliates broadcast them over the air makes the affiliates redundant and removes the last hurdle to turning the networks themselves into complete global broadcasters with the ability to draw viewers from around the world. One very early example of a decision by one of the networks to simulcast at least some of its programming online is CBS with its CBS All Access streaming service.[11] However, the service is not designed to take the place of its OTA affiliates, but to operate in conjunction with them. Nevertheless, it is easy to imagine a time very near in the future where CBS would decide that charging $5.95 a month, plus the opportunity to keep all its ad time and retransmission consent fees might just be tempting enough to drop its affiliates and go a la carte.

While a number of cable channels have also been active in delivering their programming to audiences overseas, they, like their broadcast network counterparts, will benefit from the global aspects of 21st Century Television. Like the networks, the cable channels also offer their programs online to viewers, but, like the networks, they still hold back from simulcasting their new episodes online as well as on traditional cable. However, for the cable channels, like the networks, that may be changing. HBO now offers a stand-alone streaming service called HBO Now and CBS has a stand-alone streaming service for its Showtime property. ESPN is embracing TV Everywhere on all its channels. Given that the cable industry has been traditionally more aggressive in delivering programming internationally, it may be the cable channels that will be the first to truly embrace 21ˢᵗ Century Television.

The major DTH satellite companies, DirecTV and Dish Network, have both gone international, but in a different way. While both can be received in the western hemisphere, making it possible for viewers in all the Americas potentially to become subscribers, both companies have launched international packages. These international packages allow viewers whose first language is not English to watch programming in their own language. In addition, both companies deliver programming from around the world in the international packages, thus making both companies global players, but in a reverse way from the cable channels and the networks.[12]

Additionally, both DTH satellite companies have a cooperative agreement with a company called AmericanTV2Go, which allows DirecTV and Dish Network customers the ability to watch DirecTV/Dish packages wherever they are in the world. Subscribers of AmericanTV2Go have to be a customer of DirecTV and have a DirecTV DVR or Genie system. For Dish Network, subscribers of AmericanTV2Go have to be a customer of Dish Network and have a Dish Network DVR or Hopper system. Included in the subscription to AmericanTV2Go is a Slingbox that is connected to the AmericanTV2Go facility along with the DirecTV or Dish Network DVR system, and the subscription to the AmericanTV2Go hosting service. Once the service is finalized, subscribers can watch television on their computers, iPads, etc., anywhere in the world they can get Internet access, and can even watch on a television set through their computer connection.[13] In this way, DirecTV and Dish Network are global players for their customers.

Others have been even more aggressive in making international moves. Google's YouTube has been and continues to be a global sensation, with videos from around the world available to worldwide audiences. Netflix has become an international distributor, providing its programming in 190 different countries, and worldwide, has more than 137 million subscribers as of the third quarter of 2018, 57% of which are international.[14] Amazon Prime Video/Instant Video – perhaps the major competitor to Netflix – has itself gone global and now reaches audiences in more than 200 countries, surpassing Netflix in total countries reached.[15] Online channels that are international and global abound on over-the-top set top boxes such as Roku, Amazon, and Apple TV boxes.

Going Global and Audiences

Television is global, no doubt about it. Trade shows such as MIP-TV, MIPCOM, the American Film Market, and NATPE's convention are all examples of opportunities for countries around the world to sell programming to each other.[16] Syndicators can turn programs that never even make it to U.S. television into revenue-makers by selling those programs in countries across the world where audiences enjoy them. Additionally, successful U.S. television programs are sold worldwide, with episodes sometimes airing within a week of when they aired on U.S. television. The worldwide syndication market has been and continues, at this time, to be a critical part of the financial success of U.S. television.

So why change? While global syndication has been a boon for U.S. television overall, going global opens the content providers (moving back to the 21st century term for broadcast and cable channels) to vast new audiences around the world directly, without having to work through syndication channels. Twenty-first Century Television content providers, through their ability to deliver programming worldwide using the cloud for storage and delivering programming from their websites through IPTV, will have the potential to reach almost 4.4 billion people around the world – as of March of 2019 at least – and that number will continue to grow throughout the years to come.[17] Further, because U.S. television programs sell well today around the world, there is a ready-made worldwide audience for U.S. programming. Because of that ready-made audience for U.S. television programs, the delivery of 21ˢᵗ Century Television will likely help to spur the pace of the increase in Internet deployment across the globe.

Because U.S. programming is well known worldwide through the practice of U.S. companies syndicating programs to those countries, going global gives U.S. content providers the opportunity to control the delivery of their programs to countries around the world. Going global gives the content provider the ability to bypass the scheduling decisions of the various content distributors in other countries and deliver the programming directly to their worldwide audiences such as Netflix does today. Because 21ˢᵗ Century Television is not delivered over the airwaves, but reaches viewers through the Internet in the various delivery methods, the rules that govern television programming delivered over the airwaves in other countries can be eliminated or at least minimized. Content providers will be able to deliver programs in ways that audiences worldwide can enjoy them directly, without having to watch their favorite programs when their television stations carry them, or with whatever changes might have to be made to the programs to allow them to fit into the government's view of television. Audiences worldwide will be able to enjoy the same television programs and learn of new programs that have not been available to them before through the full implementation of 21ˢᵗ Century Television.

In that way, U.S. content providers can build their worldwide brand images, build the size and the diversity of their audiences, compete directly with global content providers from other countries around the globe, and, in general, become significantly more profitable. World audiences love to watch U. S. television programming – going global gives the U.S. content providers the ability to reach those worldwide audiences at the same time they are reaching their home audiences. Twenty-first Century Television, then, makes TV Everywhere, truly TV everywhere.

Additionally, by going global, 21ˢᵗ Century Television allows U.S. viewers to enjoy their programs at any time no matter where they might be traveling in the world. Regardless of the country, U.S. viewers will have the chance to keep up with their favorite programs as new episodes become available as well as being able to watch those programs that are delivered live as they occur. As long as television viewers have the opportunity to connect to the Internet in some manner, their favorite programs will be available for them directly, without the need for additional intermediary hardware, software, or subscriptions.

Further, because 21ˢᵗ Century Television will be delivered as video-on-demand, viewers will be able to view any and all of their favorite programs as they become available, either at the time of initial airing – especially important for live events and for those who are avid fans of a particular scripted program – or at a later time, but still while outside of the United States. Going global with 21ˢᵗ Century Television also makes it possible for viewers in international locations to watch their favorite programs on their tablets and smartphones in addition to their computers, as long as they have Internet access.

As has been described in previous chapters, 21ˢᵗ Century Television makes a content provider's entire inventory available to viewers. Going global, then, makes it possible for a content provider's entire inventory of programs to be available worldwide all the time, for the entire world to enjoy again and again, on the widest range of devices. The possibilities for massive worldwide audiences enjoying the inventory of a content provider makes going global a requirement in the 21ˢᵗ Century Television world.

Going Global and Revenues

In earlier chapters in this portion of the book, there were in-depth discussions regarding the enormous potential of 21ˢᵗ Century Television for producing tremendous revenue streams through advertising[18] and ubiquitous product placement.[19] Going global with 21ˢᵗ Century Television magnifies the possibilities of those revenue streams by delivering worldwide audiences anxious to watch the advertisements they want to watch (using ATMA, Programmatic, and DAI[20]) as well as enjoying the ubiquitous product placement that would be available to them as they watch their favorite programs.

The content providers of U.S. television today – more specifically, the networks and the cable/DTH satellite channels – fill their advertising slots with commercials from companies that have nationwide appeal. When you examine those advertisers, virtually all of them are transnational corporations with products and services that reach around the globe.[21] A further example – in the *2018 BrandZ Top 100 Most Valuable Global Brands,* published by the WPP Group, 17 of the top 20 most valuable global brands for 2018 were U.S. brands, led by Google, Apple, Amazon, and Microsoft. Coming in at number five was Tencent. The other two international brands in the top 20 were Alibaba Group (#9) and SAP (#17). For the full Top 100, at least 79 of the brands were either U.S. brands or brands that had a significant presence in the U.S. (e.g, Honda, Toyota, RBC, among others).[22] Every one of those 79 out of the top 100 brands advertises on U.S. television. In reality, then, those 79 most valuable global advertisers for 2018 were both U.S. national advertisers and worldwide advertisers.

The tremendous advantage of 21st Century Television is the ability to reach audiences around the globe directly and to deliver them to advertisers. Consider for a moment how much more cost effective, and therefore, how much more valuable it would be for global advertisers to place their advertisements on 21st Century Television content providers' programs that reach audiences in countries throughout the world. Further, given the targeted nature of 21st Century Television's programmatic media buying and dynamic ad insertion both driven by aggregated targeted micro-advertising, advertisers could have the expectation to successfully reach upwards of 90% of a program's audience worldwide. While each ad would be playing potentially to a smaller audience overall due to the extremely targeted nature of ATMA, the opportunity to sell an individual ad insertion numerous times to a variety of advertisers, along with the ability to sell each ad at a premium due to a high success rate, more than offsets the smaller audience for each ad. The result, then, would be a larger total revenue for each ad insertion location, especially if each specific ad insertion was reaching the highly-targeted audience in each of as many as 200+ different countries. Such an opportunity for success would be extremely profitable for both the advertisers and the content providers.

Twenty-first Century Television's ability to deliver global audiences also brings into play the opportunity for content providers to connect with international advertisers that might not otherwise advertise on U.S. television. Remember, 21 out of the top 100 most valuable global brands for 2018 do not have a presence or do not have a significant presence on U.S. television because their target markets do not include the U.S. at this time. Given the nature of 21st Century Television, every one of those 21 top 100 brands currently not having a presence on U.S. television today would likely find it extremely profitable to advertise with the 21st Century Television content providers. They would be able to reach their audiences around the world with the powerhouse programs of the U.S. content providers, as well as introduce their products and brands to U.S. television audiences. Further, because most of those companies of the 21 top 100 brands without a presence/significant presence in the U.S. are located in China, bringing those companies onto 21st Century Television would open China's markets and its 829 million Internet users (as of March 31, 2019)[23] to viewers around the world, and advertisers from around the world to those Chinese Internet users. The reciprocal nature would provide exceptional opportunities for all concerned.

Today, for a U.S. program airing in, say, 20 different countries around the world, advertisers – both U.S. as well as international ones – have to negotiate (or more likely the advertising agency representing the advertiser would have to negotiate) potentially with each of the different television stations carrying that program in those 20 countries or even with their respective governments. To place advertising in that program in all 20 countries requires first that the advertiser or the agency have a presence in each country in some form. Second, there are likely to be different requirements for placing advertising in that program, depending on the country. Third, payment likely has to be made in several different currencies (unless the 20 countries are all in the Euro zone), requiring conversion of each currency at a fluctuating exchange rate. Fourth, depending on the country, there may be a lack of protection for the advertiser in such areas as rate increases during a contract period, make-good opportunities should an advertisement not run as scheduled, or even that the ad will not run directly against its competitors during the same break, among others. While there are numerous other considerations that would need to be negotiated, these four are representative of the potential problems with attempting to advertise on a given program when it airs in a variety of different countries. These problems and others make such advertising complex, time-consuming, and expensive.

Twenty-first Century Television's global reach eliminates all these concerns. With 21st Century Television, the advertiser contacts the content provider originating the program and negotiates placing advertising into the program for each country, thus eliminating the need for a presence in each one. Because the television program is delivered over IPTV, there will be only one set of requirements for placing the advertisement on the content provider. (Note: There may still be requirements when it comes to *airing* the advertisement in one or more of the countries, but any such requirements of that nature would be handled at the outset of negotiations and would still be between the advertiser and the content provider.) Payment is made one time, in one currency, eliminating the need for multiple currencies and multiple currency exchanges.

Protections for the advertiser would be negotiated only with the content provider because the content provider would be the only one delivering the program to the 20 countries. Other concerns and problems would also be negotiated between the content provider and the advertiser, so instead of a complex, time-consuming, and expensive multi-country process like today, 21st Century Television requires only one set of negotiations, making the process simple, much less time-consuming, and less expensive. (The process at least; the cost of the advertising could be more expensive, but the total cost of both the advertisement plus the process would be less expensive, especially when time and complexity are figured into the costs.)

Further, 21st Century Television allows advertisers from other countries the ability to advertise on a given content provider's program only in the countries where they have markets. Using the previous example, today, if an advertiser wanted to reach only ten of the 20 countries airing a particular program, the advertiser would still have the same concerns as before for each of the 10 countries. With 21ˢᵗ Century Television, the advertiser would negotiate with the content provider to air targeted ads using programmatic media buying and DAI both driven by ATMA for those ten countries only, leaving an advertising availability for one or more different advertisers to reach all or parts of the other ten countries where the availability exists. In addition to being simpler, less time-consuming, and less expensive overall, 21ˢᵗ Century Television also provides a flexibility that is simpler, less time-consuming, less expensive for the advertiser(s), and likely much more profitable for the content provider.[24]

Going global also allows 21ˢᵗ Century Television to expand its opportunities for ubiquitous product placement within the programs of the content providers. Globalization has brought and will continue to bring the world's products and services to audiences across the globe. As 21ˢᵗ Century Television programs evolve and develop, new locations, new peoples, and new cultures will continue to be introduced.[25] With that evolving program diversity will come opportunities for new and additional product placements for content providers to sell to advertisers for inclusion in their programs.

Much as with advertising, 21ˢᵗ Century Television will make the ubiquitous product placement discussed in Chapter 8 a reality and an enormous potential revenue stream currently developing in both online and legacy media programming. The ubiquitous product placement opportunities for companies all over the world will help to create new markets for companies' products. By reaching out through ubiquitous product placement, companies will be able to reach consumers anywhere in the world with their products. Those consumers, then, have the chance to learn about products they likely have never seen before. They can explore those products in depth, get excited about the products, and make purchases. Suddenly, new markets open to companies through the use of ubiquitous product placement.

In the same manner, enterprising entrepreneurs will see a product they find interesting, new, and enticing in a scene from a 21st Century Television program. When they click on the product, they are taken to the product's website where they find that the product is not one currently offered in their locations. Making the decision to contact the company that makes the product those entrepreneurs begin the process of bringing the product to their markets. This scenario can and will happen again and again as the global aspect of 21st Century Television takes hold. Going global makes product placement an ever-growing revenue stream for the content providers, and has the potential to open new markets and produce tremendous new revenues for their advertisers. It's a "win-win" scenario for all concerned.

Summary

The ease and profitability of going global provides an additional powerful reason why 21st Century Television is **the** television of the future. Netflix is the current leader in reaching out to more than 137 million subscribers worldwide with its programming. The other major online content providers – Amazon, YouTube, Yahoo, even iTunes – won't be far behind in moving to a global marketplace for their programming. Further, content providers in other countries will also take advantage of 21st Century Television's ability to deliver global audiences by reaching out across the globe with their programming as well. Those that do not quickly follow suit will be left to struggle against the inevitable flow of global programming.

Already, cable organizations such as Disney (ESPN, along with Disney programming), NBC Universal, Discovery, Turner channels, and, of course, the major cable news channels reach out to international audiences around the globe through their global or regional offerings. Some provide the programming with subtitles; others with local language voiceovers. The success of these organizations, even given the requirements that they have to work under, shows that the potential for even greater opportunities is within reach of each 21st Century Television content provider.

With the ability to reach global audiences, content providers will be able to increase their audiences and introduce new viewers to their program inventories. Additionally, the content providers will be able to build audiences for even classic television programs among those locations that may never have had the chance to see the programs before. Further, with the ability to accurately target audiences around the world with advertisements they want to watch, revenues from both advertising and product placement will soar. Analog dollars to digital pennies? No, with 21st Century Television plus global reach, it will be analog dollars to digital trillions!

PART III:
VISIONING THE FUTURE

CHAPTER 11
FINAL THOUGHTS AND VISIONS

T rue 21st Century Television is fast approaching. Already the television industries are seeing and dealing with massive changes almost on a daily basis. Mergers and acquisitions are occurring seemingly every time the consumer turns around. Local station groups are getting together creating mega-groups which threaten not only the cable, satellite, and OTT competitors, but also the networks those stations are affiliated with. ATSC 3.0 – should it actually become an alternative to other forms of Internet delivery (instead of a hoped-for future development or a bust) – will significantly impact the 21st Century Television scene and possibly negate much of what has been accomplished so far this century. Where television will be in the year 2100 is anyone's guess, but there are possibilities to consider before that time.

21st Century Television will be the "people's television." Video-on-demand, delivered through some form of Internet-protocol television will be the vehicle through which content consumers (notice the name change – "viewers" will be an antiquated term as television moves into the future) will enjoy their television. Where they will enjoy their television is one of the questions that must be considered, however. "Television set" and "screens" may also be antiquated terms, because enjoying television will be much more than just a screen, and a screen will be much more than just a television set, computer, tablet, or phablet smartphone.[1]

Instead, delivery devices (not screens) will be everywhere and anything that can deliver content to 21st Century Television consumers will deliver the content. Further, consumers will have the opportunity to enjoy their television in every way from "lean-back" passive viewing to full-on "lean-forward" active experiencing of the television content. For the passive viewer, delivery devices as screens will be everywhere. Televisions will no longer be sets as today's viewer thinks of them. Rather, imagine – as Ray Bradbury did in his futuristic world in *Fahrenheit 451* – television walls and television rooms where the viewer can retreat to and enjoy a 360-degree television version of the program. Add a floor and a ceiling television and the room becomes a truly immersive viewing opportunity.

For those not willing or unable to afford the "television viewing room," the content consumer can retreat behind the virtual reality headset – or more likely, glasses,[2] or even a "virtual reality lens" that can be placed over the eye or embedded in the eye itself and switched on or off. With the VR device, the person can escape to wherever his/her desires take her/him without having to worry about being at home in the TV room. The viewing experience can happen anywhere – which, of course, is already in its early stages of deployment. Further, with VR glasses or lenses, the consumer would have the opportunity to switch on and off from the VR viewing simply by taking the glasses off or switching the lens off.

For the traveler, television viewing will also be an anywhere, anytime enjoyment opportunity. Having to travel 500 miles by automobile? No problem – the consumer simply starts her/his autonomous car, tells the vehicle where (s)he wants to go, then sits (or lies) back, switches on the content delivery device, and the automobile becomes awash with the program(s) (s)he has chosen – front and back windshields and side windows immerse the passenger(s) in the viewing opportunity. Eight to eight-and-one-half hours later, depending on the speeds the autonomous automobile drives, the content consumer has arrived at his/her destination, having binge-watched a number of episodes of *Game of Thrones*.[3] Further, should passengers in the rear seat prefer to consume some other video program, the rear windshield and rear side windows can deliver the different viewing experience while the viewers in the front seat enjoy their episodes of *GoT*. Wireless earbuds or highly directional speakers could provide the audio to each individual consumer without having to bother the others in the automobile. Additionally, should one passenger in the front or rear seat of the automobile prefer to view the surrounding countryside for a while, the windshields and windows would return to normal for the passenger, while still delivering the program choice to the "driver" (a term used only for placement in the automobile) or other passenger(s). Alternatively, any of the travelers in the automobile could use her/his VR glasses or lenses and enjoy a visual viewing experience of his/her choosing.

When the traveler(s) reaches her/his destination, if (s)he checks into some form of accommodation, (s)he will find the opportunity to enjoy content at the location. Whether on some form of set/device or on the mirror while preparing for bed or the next day's events, the opportunity for viewing will be available. In reality, the opportunity just described already exists in some hotels – usually high-end hotels – around the country and the world.[4] These types of mirrors will also be a mainstay in homes and apartments around the world as well.

On the other end of the television spectrum, there is the "lean-forward" immersive experience of television. This television opportunity moves the content consumer (notice now why the change from "viewer" is important) from passive viewing to an active experiential form of television enjoyment. While this possibility is still likely a decade or two away, the beginnings of a "holodeck"[5] television experience is already in the early stages of being developed.[6] Imagine, then, the opportunity for a content consumer to fully enjoy the experience of the video content (s)he has in front of him/her. Through the use of holograms, whole immersive storylines are created and the consumer has the opportunity to participate fully in the television experience. Through the use of lightning fast (10+Gb/s) Internet delivery systems, fully interactive holographic content can be available for the consumer in her/his own television room for her/his enjoyment. Choose an experience of skiing down a mountain and the consumer is right there with virtual skis and poles ready at her/his decision time to begin the run, complete with the virtual feel of snow, wind, sounds, etc. The same experiential opportunities will be available with any content, scripted or unscripted.

Monetizing the Future

While the future for the consumer seems unlimited in its excitement and opportunities, money still has to be made by the content producers and content providers. Subscriptions for channels which are all the rage today simply will not produce enough revenue for the future productions, although some would like to believe in a future of all-subscription television.[7] Further, there is only so much that consumers are willing to pay for television subscriptions before the costs become as expensive as today's cable bundles.

The answer to monetization is, of course, subscriptions for some channels, but advertising and all the other aspects discussed in this book for the other channels. As 21st Century Television progresses, more and more content providers will see the importance of addressable advertising, ubiquitous product placement, endless promotion, and a global reach to generate the revenue necessary to fund this 21ˢᵗ Century "Diamond Age" of Television.

Consider, as described, the development of addressable advertising and the future. As stated much earlier, and verified by Deloitte and other studies, people are willing to watch commercials as long as they are the commercials they want to watch, for the products they like, and with the creatives they desire. Section one of Part II of this book has already shown in detail how the three-step process of aggregated targeted microadvertising, programmatic media buying, and dynamic ad insertion will make the process of reaching each consumer – or groups of consumers – with the advertising(s)he wants in the content (s)he likes to enjoy, the product (s)he uses and/or desires, and the creatives (s)he prefers easy, affordable, and extremely profitable. Combining all this together – as stated in the earlier chapters – makes the likelihood of a successful advertisement at least 80% or higher. Further, because of the interactive nature of 21ˢᵗ Century Television advertising, the 80% of the ads that are successful will virtually guarantee numerous "impulse buys" of the products being advertised.

Adding in the extra dimension of the "lean-forward" immersive television experience, makes it possible for the consumer to virtually experience the product that is of interest to him/her. Instead of only seeing a model demonstrating a product, in the immersive television experience mode, the consumer can "try out," "try on," or "use" the product virtually. Imagine being able to try on a new set of clothes, fish a stream with a fly rod, or see how the newest automobile is designed "up close and personal." The opportunities for immediate purchases are practically boundless.

The same goes for ubiquitous product placement. In the "lean-back" viewing mode, the viewer has the opportunity to select an object on any part of the viewing area for more information and purchase opportunities. As described earlier, anything selected will open a discreet window on a different portion of the viewing area (remember, at this time, the idea of a television is wall-sized or at least much larger than today's television sets). The consumer then has the opportunity to consider the product that is in the window and decide to purchase, hold, or reject the product (s)he is considering.

In the "lean-forward" immersive, interactive television experience mode, the consumer has the opportunity to actually interact with the object of her/his curiosity right there in front of him/her. The action of the program will continue around the consumer, but the consumer will have the opportunity to see, feel, and discuss the object with others in the scene who may be connected with the product (say, the owner or "driver" of a particular automobile). The consumer would then have the option to move out of the program (s)he is experiencing to a different scenario that would allow her/him to virtually ride in the automobile (it would be autonomous, after all), and a change to a third scenario could take him/her to a dealer scenario where the consumer could purchase the automobile and have it delivered to her/his home address. All this could be accomplished while still in the consumer's personal "experiential television room" with two simple voice commands for changing the scenarios (or, by mid-century, simply two thought commands).

Additionally, in both modes, the consumer has the option to move between the modes for her/his best decision-making options. So, even in the "lean-back" mode, the consumer could move into the "lean-forward" mode long enough to virtually interact with the product before returning to the "lean-back" mode to purchase or not purchase the product and return to the viewing opportunity (s)he had just left.

Further, these opportunities for purchases are not simply limited to a single country. Remember, in the chapter on the global market, the future for 21st Century Television is one of true global competition. With the viewer/consumer in charge of his/her television, countries' borders must disappear for television. Any limitations on television enjoyment by the world's population will be devastating to a country's government or television industry that demands limitations. As such, all that has been described in the previous chapters and in the previous pages of this chapter must be considered on a complete global scale by all content producers, content providers, advertising agencies, and the advertisers themselves. The content consumers – the world's television audience – will demand it, and the television industries will happily comply.

A Personal Summary

Seven years ago, I wrote my first book, the first edition of *21ˢᵗ Century Television: The Players, The Viewers, The Money*. The "Epilogue" of the book is even truer today than it was in 2012. In very few words, it encompassed the changes that were beginning at that time, those that have come to fruition by today in 2019, and the changes to look forward to in the future. I'll close with a portion of that epilogue:

21ˢᵗ Century Television is, at its core, about freedom. For the viewer, this means freedom to watch his/her favorite program in the way in which (s)he wishes to watch it. It's a freedom from the tyranny of appointment viewing. It's a freedom from schedules, from only one or two opportunities to watch a favored episode. It's a freedom from having to wait until some programmer decides to air that favored episode. It's a freedom from having to remember to set a DVR to record a program the viewer will miss because of another obligation. It's a freedom from setting a DVR for one program because there is a conflicting program. It's a freedom from the tyranny of program packages that requires a viewer to purchase 500 channels of programming when (s)he wants only 10. It's a freedom to watch advertising that makes sense for the viewer's interests, wants, and needs, instead of having to endure advertising that has no relevance to, or does not connect with, the viewer. It's a freedom to watch a program on the go, unencumbered by time, location, or lack of a television set. Ultimately, for the viewer, 21st Century Television is about the freedom to enjoy television – all of television – on his/her time schedule, when (s)he wants to watch, where (s)he wants to watch, on whatever platform is most convenient or preferred by the viewer.

But it is also freedom for the television industry. For the industry it is a freedom from programming schedules and programming strategies. It's a freedom from the overnights, from the sweeps. It's a freedom to connect with the audience through promotional techniques that allow each content provider to truly understand who the audience is for every program, right down to each individual in that audience. It's a freedom to deliver programming that is right for each viewer, that each viewer wants to enjoy, and so can become loyal to that program because it is right for her/him. It's a freedom to make money – lots of money – even more money than the revenues of today - through advertising that works and works for everyone – the viewer, the content provider, and the advertiser.

Finally, it's a freedom from boundaries - local, regional, national, and even international boundaries. It's a freedom to reach audiences wherever they are around the world, to deliver the programming and advertising that works for them, and – at the same time – enhance the reputation as well as the bottom line of each content provider and the television industry as a whole.[8]

Enjoy this ride – it will be amazing, exhilarating, frustrating, and ultimately, extremely rewarding for all parties involved.

NOTES

Introduction

1. Zucker, Jeff, "Keynote Address," 2008 National Association of Television Program Executives convention, January 28-31, 2008, Las Vegas, NV.
2. ibid.
3. "Now it's digital quarters for analogue dollars," *Videonet*, (March 3, 2011), http://www.v-net.tv/now-it%E2%80%99s-digital-quarters-for-analogue-dollars/
4. See "TW Cable Launches Interactive Ads," *LightReading–Cable*, (March 8, 2012), http://www.lightreading.com/document. asp?doc_id=218532&site=lr_cable and Kaufman, Alexander C., "PayPal Teams With TiVo, Comcast for Interactive Click-And-Buy Ad Service," *The Wrap*, (June 12, 2012), http://www.thewrap.com/media/article/paypal-teams-tivo-comcast-interactive-click-and-buy-ad-service-43896
5. "Cablevision Unveils Optimum Select," *PRWeb*, (September 16, 2009), http://www.prweb.com/releases/2009/09/prweb2890 344.htm
6. See the section on Connected TVs in "Chapter 7 – Mobile DTV, Connected TV, and the iWorld" and "Chapter 6 – Over-The-Top Set Top Boxes" in Aycock, Frank A. (2015),

21st Century Television: The Players, The Viewers, The Money, 2nd ed., (Charleston, SC: CreateSpace).

7. For an in-depth discussion of Aggregated Targeted Microadvertising, see Chapter 5.
8. "Dynamic Ad Insertion," *IAB Wiki,* http://www.iab.net/wiki/index.php/Dynamic_ad_insertion.
9. "Dynamic Ad Insertion," *itv disctionary,* http://www.itvdictionary.com/definitions/dynamic_targeted_ad_insertion_definition.html.
10. *Wall Street Journal,* (February 4, 2010).

Chapter 1 – History of Advertising

1. For example, the line item for radio and television in the Bulgarian federal budget.
2. For example, the tax that is collected in the Netherlands.
3. One only has to look at the development of the television advertisement in any of the Central European countries after the move away from communism in the late 1980s and early 1990s to see the speed at which the advertising changed from 60 second ads to 30 second ads. The combination of the need to compete with European and U.S. advertising airing on satellite channels such as CNN International, and the technological changes in advertising and television generally, literally forced the public to develop their sophistication for understanding the 30 second commercial almost overnight.
4. Head, Sydney W and Sterling, Christopher H., *Broadcasting in America,* (Boston: Houghton Mifflin), c. 1990, pg. 61.
5. Gross, Lynne Schafer, *Telecommunications: An Introduction to Radio, Television and the Developing Media,* (Dubuque, Iowa: Wm. C. Brown Company), c. 1983, p.77.
6. ibid.

7. " History: 1950s – AdAge Encyclopedia of Advertising," *AdAge*, (September 15, 2003), https://adage.com/article/adage-encyclopedia/history-1950s/98701/

8. ibid.

9. Braswell, Sean, "The Cigarette Company That Reinvented Television News," *Ozy*, (February 22, 2019), https://www.ozy.com/flashback/the-cigarette-company-that-reinvented-television-news/92498.

10. ibid.

11. "History: 1950s…," loc. cit.

12. Mitchell, Chase, "Character Study: Speedy Alka Seltzer," *MLT Creative*, https://www.mltcreative.com/blog/character-study-speedy-alka-seltzer/

13. Pathak, Shareen, "Yabba Dabba Cough! Flashback to When the Flintstones Shilled Cigarettes," *AdAge*, (April 2, 2013), https://adage.com/article/rewind/yabba-dabba-cough-flintstones-shilled-cigarettes/240572/

14. Blakemore, Erin, "The Rigged Quiz Shows That Gave Birth to 'Jeopardy!'" *History.com*, (March 29, 2019), https://www.history.com/news/quiz-show-scandal-fraud-jeopardy

15. ibid.

16. "Weaver, Sylvester L.," *AdAge*, (September 15, 2003), https://adage.com/article/adage-encyclopedia/weaver-sylvester-l/98932/

17. "1960s Creativity and Breaking the Rules," *AdAge*, (March 28, 2005), https://adage.com/article/75-years-of-ideas/1960s-creativity-breaking-rules/102704/

18. ibid.

19. "History 1970s | AdAge Encyclopedia of Advertising," *Ad Age*, http://adage.com/article/adage-encyclopedia/history-1970s/98703/

20. ibid.

21. "Television | AdAge Encyclopedia of Advertising," *Ad Age,* http://adage.com/article/adage-encyclopedia/television/98901/
22. ibid.
23. ibid. The remedy for this is the focus of this book, and will be discussed in detail in the coming chapters.
24. For instance, the cost of a 30-second ad in the 2018 Super Bowl telecast was slightly more than $5 million. (Source: Carroll, Charlotte, "Super Bowl LII: How Much Does an[sic] Commercial Cost," *Sports Illustrated,* January 11, 2018, https://www.si.com/nfl/2018/01/11/super-bowl-lii-ad-cost
25. "U.S. TV Ad Spending to Fall in 2018," *eMarketer,* March 28, 2018, https://www.emarketer.com/content/us-tv-ad-spending-to-fall-in-2018.

Chapter 2 – The Television Players

1. See "Chapter 2 – The Cable Television Industry" in Aycock, Frank A. (2015), *21ˢᵗ Century Television: The Players, The Viewers, The Money,* 2ⁿᵈ ed., (Charleston, SC: CreateSpace).
2. They were ABC, CBS, and NBC.
3. Aycock, Frank A. (1989), *A Comparison of Blunting, Hybrid, and Counterprogramming Television Strategies and Their Effects on Total Network Television Viewing.* Doctoral Dissertation.
4. Gross, Lynne Schafer, *Telecommunications: An Introduction to Radio, Television and the Developing Media,* (Dubuque, Iowa: Wm. C. Brown Company), c. 1983, p.77.
5. ibid.
6. A.C. Nielsen Co. (2019), *The Nielsen Total Audience Report, Q3 2018,* https://www.nielsen.com/content/dam/corporate/us/en/reports-downloads/2019-reports/q3-2018-total-audience-report.pdf
7. See "Chapter 2 – The Cable Television Industry," Chapter 3 – The DTH Satellite Industry," and "Chapter 4 – Internet

Protocol Television" in Aycock, Frank A. (2015), *21ˢᵗ Century Television: The Players, The Viewers, The Money*, 2nd ed., (Charleston, SC: CreateSpace) for a deeper discussion.

8. See, for example, FCC, *Incentive Auction FAQs*, https://www.fcc.gov/general/incentive-auctions-faqs, and nab, *Spectrum Repacking*, https://www.nab.org/repacking/

9. "Video Compression Standards–Pros&Cons," *Mistral*, https://www.mistralsolutions.com/articles/video-compression-standards-pros-cons/

10. Aycock, Frank A, "Millennials and Gen Zs: Broadcasters, ATSC 3.0, and the Death of Netflix!" *LinkedIn,* (April 9, 2019), https://www.linkedin.com/pulse/millennials-gen-zs-broadcasters-atsc-30-death-netflix-frank-aycock/

11. ibid.

12. For a deeper discussion of this subject, see "Chapter 7 – Mobile DTV/ATSC 3.0, Connected TV, and the iWorld" in Aycock, Frank A. (2015), *21ˢᵗ Century Television: The Players, The Viewers, The Money*, 2nd ed., (Charleston, SC: CreateSpace).

13. Subscription Video-on-Demand

14. Advertising (or Ad-based) Video-on-Demand

15. Arbel, Tali, "Disney takes over Hulu from Comcast as stream wars heat up," *AP News,* (May 14, 2019), https://www.apnews.com/bc8fe3077bc5414e979436f9b596da30

16. Moraski, Lauren, "CBS launches expansive digital subscription service," *CBS News,* (October 16, 2014), https://www.cbsnews.com/news/cbs-launches-digital-subscription-service-cbs-all-access/

17. Gross, Lynne Schafer, *Telecommunications: An Introduction to Radio, Television and the Developing Media,* (Dubuque, Iowa: Wm. C. Brown Company), c. 1983

18. ibid.

19. IPTV - Internet Protocol Television

20. Gross, op. cit.

21. See "Chapter 2 – The Cable Television Industry" in Aycock, Frank A. (2015), *21ˢᵗ Century Television: The Players, The Viewers, The Money*, 2ⁿᵈ ed., (Charleston, SC: CreateSpace).

22. A.C. Nielsen Co., op. cit.

23. See "Chapter 2 – The Cable Television Industry" in Aycock, Frank A. (2015), *21ˢᵗ Century Television: The Players, The Viewers, The Money*, 2ⁿᵈ ed., (Charleston, SC: CreateSpace).

24. See "Chapter 15 –Advertising" in Aycock, Frank A. (2015), *21ˢᵗ Century Television: The Players, The Viewers, The Money*, 2ⁿᵈ ed., (Charleston, SC: CreateSpace).

25. See "Chapter 18 – Promotion" in Aycock, Frank A. (2015), *21ˢᵗ Century Television: The Players, The Viewers, The Money*, 2ⁿᵈ ed., (Charleston, SC: CreateSpace).

26. For more information on what a "skinny bundle" is, see Lovely, Stephen, "What Is a 'Skinny Bundle'?" *CordCutting*, (January 24, 2019), https://cordcutting.com/blog/what-is-a-skinny-bundle/

27. Segan, Sascha, "What Is 5G?" *PCMag*, (April 16, 2019), https://www.pcmag.com/article/345387/what-is-5g

28. Gruenwedel, Erik, "Study: Satellite TV Drives 47% Surge in Pay-TV Subscriber Losses in 2018," *MediaPlayNews*, (March 7, 2019), https://www.mediaplaynews.com/study-satellite-tv-drives-47-surge-in-pay-tv-subscriber-losses-in-2018/

29. Forrester, Chris, "It Was 20 Years Ago Today…", *SatMagazine*, (February 5, 2009), http://www.satmagazine.com/story.php?number=1053209847

30. Horowitz, Jeremy, "Dish Network considers $10 billion 5G network instead of spectrum sale," *VentureBeat*, (October 8, 2019), https://venturebeat.com/2018/10/08/dish-network-considers-10-billion-5g-network-instead-of-spectrum-sale/

31. See "Chapter 4 – Internet Protocol Television" in Aycock, Frank A. (2015), *21ˢᵗ Century Television: The Players, The Viewers, The Money*, 2ⁿᵈ ed., (Charleston, SC: CreateSpace).

32. For a deeper discussion of this subject, see "Chapter 4 – Internet Protocol Television" and "Chapter 7 – Mobile DTV/ ATSC 3.0, Connected TV, and the iWorld" in Aycock, Frank A. (2015), *21ˢᵗ Century Television: The Players, The Viewers, The Money*, 2ⁿᵈ ed., (Charleston, SC: CreateSpace).

Chapter 3 – The Television Viewers

1. Weston, Miles, "Content Insider #148 – The iGen," *Digital Producer Magazine*, (August 31, 2010), http://digitalproducer. digitalmedianet.com/articles/viewarticle.jsp?id=1192752-0

2. Deloitte's State of the Media Democracy Survey, Sixth Edition, February 2012, Deloitte Development LLC, www. deloitte.com/us/mediademocracy

3. http://geography.about.com/od/populationgeography/a/ babyboom.htm

4. ibid.

5. http://www.u-s-history.com/pages/h2061.html

6. ibid.

7. ibid.

8. For examples, see early Bugs Bunny, Road Runner, Mighty Mouse, and Popeye cartoons. Also see "Chapter 1 – The Broadcast Industry" in Aycock, Frank A. (2015), *21ˢᵗ Century Television: The Players, The Viewers, The Money*, 2ⁿᵈ ed., (Charleston, SC: CreateSpace).

9. See "Chapter 2 – The Cable Industry" in Aycock, Frank A. (2015), *21ˢᵗ Century Television: The Players, The Viewers, The Money*, 2ⁿᵈ ed., (Charleston, SC: CreateSpace).

10. http://www.infoplease.com/ipa/A0005067.html

11. http://legalcareers.about.com/od/practicetips/a/ GenerationX.htm

12. http://apps.americanbar.org/lpm/lpt/articles/mgt08044. html

13. Keeter, Scott and Paul Taylor, "The Millennials," *Pew Research Center,* (December 11, 2009), http://pewresearch. org/pubs/1437/Millennials-profile

14. As a personal example, my 28-year-old son routinely sends between 1,500 – 2,000 text messages a month, and receives approximately the same number as he sends. He is a typical millennial [granted, a father's comment] who is either sending or receiving a total of at least 100 text messages a day. According to my son – yes, it is only anecdotal, I admit, but interesting none-the-less – many of his female friends will double or even triple his daily text messaging level. As a Baby-Boomer who does text some, it makes me wonder how he and his friends get anything done!

15. Keeter and Taylor, loc. cit.

16. Vasquez, Diego, "For Millennials, TV just isn't that important," *Media Life Research,* (December 1, 2011), http:// www.medialifemagazine.com:8080/artman2/publish/ Research_25/For-millenials-TV-just-isn-t-that-important. asp

17. Weston, Mike, "Content Insider #148 – The iGen," *Digital Producer Magazine,* (August 31, 2010), http://digitalproducer. digitalmedianet.com/articles/viewarticle.jsp?id=1192752-0

18. Aycock, Frank A., *21ˢᵗ Century Television: the Players, The Viewers, The Money*

19. See, for instance, the most recent Nielsen *Total Audience Surveys* and the *Digital vs Traditional Media Consumption* study by Global Web Index.

20. Clark, Travis, "Your favorite Netflix shows are more likely to be canceled after 2 or 3 seasons than those on traditional TV," *Business Insider,* (April 9, 2019), https://www. businessinsider.com/netflix-cancels-many-tv-shows-after-2-or-3-seasons-report-2019-4

21. For a more in-depth discussion of the future of 21ˢᵗ Century Television, please see the final chapter in Aycock, Frank A

(2015), *21ˢᵗ Century Television: The Players, The Viewers, The Money*, 2ⁿᵈ ed, (Charleston, SC: CreateSpace).

22. This was a phrase used by President John F. Kennedy in his inaugural address, January 20, 1961.

Chapter 4 – Advertising Overview

1. See Chapter 1 – History of TV Advertising
2. See "Chapter 17 – Retransmission Consent Fees" in Aycock, Frank A. (2015), *21ˢᵗ Century Television: The Players, The Viewers, The Money*, 2ⁿᵈ ed., (Charleston, SC: CreateSpace).
3. Baumgartner, Jeff, "Amazon Joins OTT's Big Free-for-All," *Light Reading*, (January 11, 2019), https://www.lightreading.com/video/ott/amazon-joins-otts-big-free-for-all/a/d-id/748764
4. See "TW Cable Launches Interactive Ads," *Light Reading –Cable*, (March 8, 2012), http://www.lightreading.com/document.asp?doc_id=218532&site=lr_cable and Kaufman, Alexander C., "PayPal Teams With TiVo, Comcast for Interactive Click-And-Buy Ad Service," *The Wrap*, (June 12, 2012), http://www.thewrap.com/media/article/paypal-teams-tivo-comcast-interactive-clickand-buy-ad-service-43896
5. "Cablevision Unveils Optimum Select," *PRWeb*, (September 16, 2009), http://www.prweb.com/releases/2009/09/prweb2890344.htm
6. Moran, Chuck, "The future of TV advertising in today's digital world," *MARTECH Today*, https://martechtoday.com/the-future-of-tv-advertising-in-todays-digital-world-215750
7. See the section on Connected TVs in "Chapter 7 – Mobile DTV, Connected TV, and the iWorld" in Aycock, Frank A. (2015), *21ˢᵗ Century Television: The Players, The Viewers, The Money*, 2ⁿᵈ ed., (Charleston, SC: CreateSpace).
8. "Addressable TV Advertising: Creating a Better, More Personal TV and Video Experience," *Think with Google*, June 2016,

https://www.thinkwithgoogle.com/marketing-resources/addressable-tv-advertising-personal-video-experience/

9. "Addressable TV Advertising," *IT Glossary – Gartner Research,* https://www.gartner.com/it-glossary/addressable-tv-advertising

10. "Dynamic Ad Insertion," *IAB Wiki,* ttp://www.iab.net/wiki/index.php/Dynamic_ad_insertion.

11. "Dynamic Ad Insertion," *itv dictionary,* http://www.itvdictionary.com/definitions/dynamic_targeted_ad_insertion_definition.html.

Chapter 5 – Aggregated Targeted Microadvertising

1. See "Chapter 17 – Retransmission Consent Fees" in Aycock, Frank A. (2015), *21st Century Television: The Players, The Viewers, The Money,* 2nd ed., (Charleston, SC: CreateSpace).

2. Wall Street Journal, (February 4, 2010).

3. See Chapter 9 – Promotion.

4. See Chapter 10 – Monetizing the Global Market.

5. See Chapter 9 – Promotion.

6. "John Wanamaker," *Advertising Age,* (March 29, 1999), http://adage.com/article/special-report-the-advertising-century/john-wanamaker/140185/

7. Anderson, Janna and Lee Rainie, "Future of the Internet IV," *Pew Internet & American Life Project,* (February 19, 2010), http://www.pewinternet.org/Reports/2010/Future-of-the-Internet-IV.aspx

8. ibid.

9. ibid.

10. ibid.

11. Boris, Cynthia, "'Online Privacy is Dead' Says Study and Millennials Are Okay with That," *Marketing Pilgrim,* (April 22, 2013), http://www.marketingpilgrim.com/2013/04/

online-privacy-is-dead-says-study-and-Millennials-are-
okay-with-that.html

12. "Is online privacy over? Findings from the USC Annenberg
Center for the Digital Future show Millennials embrace a
new online reality," *USC Annenberg – News,* (April 22, 2013),
http://annenberg.usc.edu/news/around-usc-annenberg/
online-privacy-over-findings-usc-annenberg-center-digital-
future-show

Chapter 6 - Programmatic Advertising

1. Marshall, Jack, "WTF is programmatic advertising?"
Digiday, (February 20, 2014), https://digiday.com/media/
what-is-programmatic-advertising/
2. Blattberg, Eric, "WTF is programmatic TV advertising?"
Digiday, (October 13, 2014), http://digiday.com/platforms/
wtf-programmatic-tv-advertising/
3. ibid.
4. ibid.
5. Locations in this case can refer not only to a city or state,
but also a region, or even any country around the world
where the viewer(s) can be reached.
6. Content providers can be networks and cable channels
from the legacy media, as well as all the new media channels (OTT, subscription, YouTube-style, etc.)
7. Unless otherwise cited, the statements throughout the
list are either taken directly from the following article or
are the author's own comments. Generally, the author's
comments are designed to update those stated by
Kantrowicz, Alex, "10 Things You Need to Know Now About
Programmatic Buying," *AdAge,* (June 1, 2015), https://adage.
com/article/print-edition/10-things-programmatic-
buying/298811

8. "65% of digital media to be programmatic in 2019," *Zenith Media,* (November 19, 2018), https://www.zenithmedia.com/65-of-digital-media-to-be-programmatic-in-2019/ and Farveen, Farzanah, "Ad dollars: How much are marketers putting aside for programmatic in 2019? *Marketing,* (November 19, 2018), https://www.marketing-interactive.com/ad-dollars-how-much-are-marketers-putting-aside-for-programmatic-in-2019/

9. Friedman, Wayne, "TV Stations' Consortium Launches Programmatic TV Ad Effort," *MediaPost,* (September, 26, 2018), https://www.mediapost.com/publications/article/325656/tv-stations-consortium-launches-programmatic-tv-a.html

10. Friedman, Wayne, "Correction: Videa In 150 Markets," *MediaPost,* (September 27, 2018), https://www.mediapost.com/publications/article/325746/correction-videa-in-150-markets.html

11. Benes, Ross, "For Most Brands, Biggest Benefit to Using In-House Is Cost," *eMarketer,* (October 23, 2018), https://www.emarketer.com/content/for-most-brands-biggest-benefit-to-using-in-house-is-cost

12. Gorman, Mark, "Brands are moving to in-house programmatic: Is it the right choice?" *ClickZ,* (October 11, 2018), https://www.clickz.com/in-house-programmatic-right-choice

13. All the information in this section is quoted, with thanks, from Aguilhar, Ligia, "The differences between AI, machine learning, programmatic buying and deep learning," *strikesocial,* (n.d.) https://strikesocial.com/blog/the-difference-between-ai-machine-learning-programmatic-advertising-deep-learning/

14. All the information in this section is quoted, with thanks, from Elkin, Tobi, "How Artificial Intelligence Ties Into Programmatic Media," *MediaPost,* (December 29, 2016), https://www.mediapost.com/publications/article/291939/how-artificial-intelligence-ties-into-programmatic.html

Chapter 7 - Dynamic Ad Insertion

1. "Glossary, Dynamic Ad Insertion," *TubeMogul*, https://www.tubemogul.com/glossary/dynamic-ad-insertion/
2. "Canoe Rides Growth Wave In 2014," *Multichannel News*, (February 4, 2015), http://www.multichannel.com/news/technology/canoe-rides-growth-wave-2014/387622#.VNOtPAxb_Vo.twitter and Saboo, Alok, "Canoe Releases Their 2014 Video on Demand (VOD) Viewed Ad Impression 2014 Year-End Report," *Fierce Cable*, (February 4, 2015), http://www.fiercecable.com/press-releases/canoe-releases-their-2014-video-demand-VOD-viewed-ad-impression-2014-year-e
3. Hamlin Media, "Video on Demand and Dynamic Ad Insertion," *Hamlin Media*, (April 16, 2015), https://www.harmelin.com/media-magnified/video-on-demand-and-dynamic-ad-insertion/
4. These results have been developed by the author after extensive research on the subject for more than 6 years.
5. Aycock, Frank A., "Chapter 15: Advertising," *21ˢᵗ Century Television: The Players, The Viewers, The Money*, 2ⁿᵈ ed, (Charleston, SC: CreateSpace), c. 2015.

Chapter 8 – Product Placement

1. *BusinessDictionary.com*, http://www.businessdictionary.com/definition/product-placement.html
2. Lafayette, Jon, "Product Placement Revenue Climbing 13.2% This Year," *Broadcasting and Cable*, (June 15, 2015), https://www.broadcastingcable.com/news/product-placement-revenue-climbing-132-year-141746
3. Mandese, Joe, "Product Placement Poised to Top $10 Billion: Reports Cites Fewer, More Valuable Deals Including Upfronts, Netflix," *MediaPost*, (June 13, 2018), https://www.

mediapost.com/publications/article/320675/product-placement-poised-to-top-10-billion-repor.html

4. *Harrison's Reports,* (January 17, 1920), pg. 9.
5. For a good listing of movies that have made use of product placement throughout movie history, see http://en.wikipedia.org/wiki/Product_placement#cite_note-12
6. "Children's Educational Television," *FCC Guide,* http://www.fcc.gov/guides/childrens-educational-television
7. Mueller, Barbara (2004), *Dynamics of International Advertising,* (New York: Peter Lang), 284-285.
8. Neer, Katherine, "How Product Placement Works," *howstuffworks,* http://money.howstuffworks.com/product-placement.htm
9. This comes from personal experience. In 2016, I had the opportunity to work in Gwangju, ROK, for a month during the summer. Being a baseball fan, I had the opportunity to see a number of games in person and on television because every game of every team was broadcast on one of the nationwide channels.
10. The two programs are known for their extensive use of product placement. Generally, they are the two programs on television that have the largest number of product placement mentions each season.
11. Buss, Dale, "'Escape Routes' Reality TV Show Opens Strong, as Ford See It," *Forbes,* (April 2, 2012), http://www.forbes.com/sites/dalebuss/2012/04/02/escape-routes-reality-tv-opens-strong-as-ford-sees-it/
12. ibid.
13. ibid.
14. ibid.
15. "Ford Motor Company New TV show 'Escape Routes'," *Autos In Transit,* (March 28, 2012), http://www.autosintransit.com/ford-motor-company-new-tv-show-escape-routes/

16. "Agenda Setting Theory," *International Agenda Setting Conference,* http://www.agendasetting.com/index.php/agenda-setting-theory

17. ibid.

18. Lazarsfeld, Paul F., Berelson, Bernard, and Gaudet, Hazel, *The People's Choice: How the Voter Makes Up His Mind in a Presendential Election,* in Lowery, Shearon and DeFleur, Melvin L., *Milestones in Mass Communication Research,* (New York: Longman, 1983), 85-112.

19. Katz, Elihu and Lazarsfeld, Paul F., *Personal Influence: The Part Played by People in the Flow of Mass Communication,* in Lowery, Shearon and DeFleur, Melvin L., *Milestones in Mass Communication Research,* (New York: Longman, 1983), 177-203.

20. "Agenda Setting theory, op. cit.

21. Sutherland, Max, "Why Product Placement works," http://www.sutherlandsurvey.com/Columns_Papers/Why%20Product%20Placement%20Works_Feb05.pdf

22. "Product placement works on reality TV, says report," *Public Citizen's Commercial Alert,* (August 22, 2011), http://www.commercialalert.org/news/archive/2011/08/product-placement-works-on-reality-tv-says-report

23. ibid.

24. ibid.

Chapter 9 – Promotion

1. Deloitte Corporation, *Digital Democracy Survey: A multi-generational view of consumer technology, media and telecom trends,* 9ᵗʰ ed., (2014), www.deloitte.com/us/tmttrends.

2. For more information, see Chapter 19 – Going Global TV in Aycock, Frank A. (2015) *21ˢᵗ Century Television: The Players, The Viewers, The Money,* 2ⁿᵈ ed., (Charleston, SC: CreateSpace)

3. YouTube Brand Channels, http://static.googleusercontent. com/media/www.youtube.com/en//yt/advertise/medias/ pdfs/brand-channel-onesheeter-en.pdf
4. Barr, Merrill, "Is YouTube Preparing for a Netflix-Style Original Content Push?" *Forbes,* (November 20, 2014), http://www.forbes.com/sites/merrillbarr/2014/11/20/ is-youtube-preparing-for-a-netflix-style-original-content- push/

Chapter 10 – Monetizing the Global TV Market

1. "CNN International," *Wikipedia,* http://en.wikipedia.org/ wiki/CNN_International#History
2. For instance, pan-European channels would include the channels of Sky Television, EuroNews and EuroSports, among others, while a good example of pan-Asian channels would be Star Television. Additionally, in Europe, viewers can see a mixture of the various countries' television programming, so that the French can watch German TV and vice versa, they both can watch British television and vice versa, etc.
3. "Internet Usage Statistics: The Big Picture," *Internet World Stats,* (March 31, 2019), http://www.internetworldstats. com/stats.htm
4. ibid.
5. Aycock, Frank A., "NAB Is Underway!" *LinkedIn,*"(April 7, 2018), https://www.linkedin.com/pulse/nab-2018-underway- frank-aycock/
6. Jessell, Harry A., "FCC Need To give OVDs MPVD Status," *TVNewsCheck,* (May 18, 2012), http://www.tvnewscheck.com/ article/59567/fcc-needs-to-give-ovds-mpvd-status?utm_ source=Listrak&utm_medium=Email&utm_term=FCC+N eeds+To+Give+OVDs+MPVD+Status&utm_campaign=Jess ell%3a+FCC+Needs+To+Give+OVDs+MPVD+Status
7. ibid.

8. ibid.
9. "Where is Netflix available?" *Netflix Help Center,* https://help.netflix.com/en/node/14164
10. See Chapter 7 - Mobile DTV, Connected TV, and the iWorld in Aycock, Frank A. (2015) *21ˢᵗ Century Television: The Players, The Viewers, The Money,* 2ⁿᵈ ed., (Charleston, SC: CreateSpace)
11. See Chapter 1 – The Broadcasters in Aycock, Frank A. (2015) *21ˢᵗ Century Television: The Players, The Viewers, The Money,* 2ⁿᵈ ed., (Charleston, SC: CreateSpace)
12. For more information on the international packages offered by DirecTV and Dish Network, go to www.directv.com or www.dish.com and search for their international packages.
13. "What We Do For You," *AmericanTV2Go,* http://www.americantv2go.com/Basics/WhatWeDo.aspx?&Hash=eb44a996-6725-49fd-8ee7-20d1e620083a
14. Iqbal, Mansoor, "Netflix Revenue and Usage Statistics (2018)," *BusinessofApps,* (November 7, 2018), http://www.businessofapps.com/data/netflix-statistics/
15. Barraclough, Leo, "Amazon Prime Video Goes Global: Available in More Than 200 Territories," *Variety,* (December 14, 2016), https://variety.com/2016/digital/global/amazon-prime-video-now-available-in-more-than-200-countries-1201941818/
16. "International Television Program Markets," *MBC – The Museum of Broadcast Communications,* http://www.museum.tv/eotvsection.php?entrycode=internationalt
17. "Internet Usage Statistics: The Big Picture," op. cit.
18. See Chapter 15 – Advertising in in Aycock, Frank A. (2015) *21ˢᵗ Century Television: The Players, The Viewers, The Money,* 2ⁿᵈ ed., (Charleston, SC: CreateSpace)
19. See Chapter 16 - Product Placement in Aycock, Frank A. (2015) *21ˢᵗ Century Television: The Players, The Viewers, The Money,* 2ⁿᵈ ed., (Charleston, SC: CreateSpace)

20. ATMA - Aggregated Targeted MicroAdvertising.
21. For example, see http://www.businessinsider.com/companies-that-spend-the-most-on-advertising-2011-6
22. WPP, *2012 BrandZ Top 100 Most Valuable Global Brands,* http://www.sayitsocial.com/wp-content/uploads/2012/05/brandz_2012_top_100.pdf
23. "China," *Internet World Stats,* (June 30, 2018), ttps://www.internetworldstats.com/stats3.htm
24. See Chapters 4-7 for an extended example of the entire addressable advertising process, including global advertising
25. Two examples would be the way the "Survivor" and the "House Hunters International" television series have made U.S. audiences aware of locations that otherwise might never have been thought of by the typical television viewer.

Chapter 11 -Final Thoughts and Visions

1. A phablet is any mobile phone, generally a smartphone, that has a screen larger than 5" – and even that is quaint these days and certainly will be in the future.
2. Think stylish sunglasses or regular frames, not Google Glass Project AR glasses.
3. *Game of Thrones* is extremely popular series that aired for eight seasons on HBO. It ended in May 2019.
4. The author has actually had the opportunity to enjoy watching television on the bathroom mirror in hotels in Washington, DC, and Warsaw, Poland.
5. For a frame of reference, consider the holodeck on *Star Trek: The Next Generation.*
6. For more information, please see the Immersive Digital Experiences Alliance (IDEA) website: https://immersiveal-liance.org/

7. A good example is my good friend, Jim O'Neill, of Brightcove and *Videomind* blog. Jim is a great person and a pre-eminent futurist on the subject of television, but he and I completely disagree on whether or not the future is all-subscription-TV or not. We'll see who's right and who's wrong, I guess, as the future plays out!

8. Aycock, Frank A. (2012), "Epilogue," *21st Century Television: The Players, The Viewers, The Money* (Charleston, SC: CreateSpace).

INDEX

Made in the USA
Columbia, SC
24 July 2019